The Diaries of a Bonedigger

Harold Rogers Wanless ·
Emmett Evanoff

The Diaries
of a Bonedigger

Harold Rollin Wanless in the White
River Badlands of South Dakota,
1920–1922

 Springer

Harold Rogers Wanless
Department of Geography and
Sustainable Development
University of Miami
Coral Gables, FL, USA

Emmett Evanoff
Department of Earth and Atmospheric
Sciences
University of Northern Colorado
Greeley, CO, USA

ISBN 978-3-031-25117-7   ISBN 978-3-031-25118-4   (eBook)
https://doi.org/10.1007/978-3-031-25118-4

This Springer imprint is published by the registered company Springer Nature Switzerland AG
The registered company address is: Gewerbestrasse 11, 6330 Cham, Switzerland

A view from near Cedar Pass looking east along the edge of the Wall of the Badlands. The hollow in the foreground is occupied by a small temporary lake formed by the blocking of one of the canyons by a large landslide. Cottonwood trees have become quite well established around this, as well as cedars, and considerable smaller vegetation. The Matterhorn-like butte to the right is capped by the *Leptauchenia* clays. (This favorite autochrome photograph of Harold Rollin Wanless was taken on July 3, 1920)

Preface

This book provides an intimate first-hand look at what the White River Badlands and its people were like and how geologists worked in the field a century ago, through the writing and photography of Harold Rollin Wanless in 1920–1923. Physically it is essentially unchanged. It is a dusty, dry place commonly with searing heat in the summer. These Badlands are also one of the most spectacular landscapes in North America, with castellated buttes and spires rising out of the prairie and winding their way for over 50 miles as a near-continuous wall. A hundred years ago the land was completely home-steaded, but it is now protected as Badlands National Park.

The rocks of the Badlands contain perhaps the most prolific fossil record of ancient mammals in North America. These fossils have been the focus of early explorers and then paleontologists for almost 175 years. This book records the gathering and preserving of that record as modern vertebrate pale-ontology was emerging.

In human ways, the White River Badlands a century ago was quite differ-ent from today. A century ago, there were many more people in the area, as every 40 acre lot was claimed as homesteads by hardworking and hopeful ranch families. The Milwaukee Railroad had arrived only a decade ago pro-viding access for agriculture and livestock to markets in Rapid City or Chicago. Also, 1920 was the time of the beginning of automobile transporta-tion, with Model T Fords navigating the rutted wagon roads. Car and espe-cially truck transportation was much slower in 1920, and the new railroads were still the main link to the outside world.

The native Lakota Sioux still held a presence in the area but were adapting to the new ways of life. 1920 was only 30 years after the Wounded Knee mas-sacre and the subsequent breaking up of the large Sioux Reservation lands that had been held in trust for the tribes. These changed their lives forever.

Secondly, this book focuses on the question of the qualities of Harold Rollin Wanless and his exposure that guided his transition from a capable student of geology to a renowned field and academic geologist. The student must have talent, abilities, and a love of nature. The student must be able to adapt to new environments and situations as fieldwork requires. Solid train-ing at a good university or college prepares the student for their career. Finally, the student should have the chance to work in the field with an estab-lished field geologist who acts as a mentor and instills intellectual curiosity for the larger meaning of what they are observing. Harold Rollin Wanless had all these attributes and advantages, and his experiences in the Badlands from 1920 through 1923 changed his life.

The two authors of this book became interactive as Emmet Evanoff, who has studied the geology of the Badlands for over a quarter century, has been trying to accurately locate the geologic features that Harold Rollin Wanless had measured and described in the 1920–22 field seasons. Harold Rogers Wanless (his son) was able to provide many photographs and a copy of the 1920 journal/diary (Chaps. 3, 4, 5, 6, 7, 8, 9, 10, 11, 12, 13, 14, 15, and 16 in this book) which began a long and rewarding interaction.

Harold Rogers Wanless was born in 1942, over 20 years after these field adventures. He is honored to be a part of presenting this look of his father's early efforts in the summer of 1920. He appreciates the love of his life, Lynn Bauer, for her support and patience in this endeavor.

Most history books are written by contemporary people trying to figure out what people did and thought and how they lived back then. This following narrative is word for word his observations, activities, and feelings from 1920, over one hundred years ago. From his numerous college-days' letters to his naturalist mother and colleagues, Harold had developed a habit of thorough observation, recording, and communication, and we now all benefit from it.

As with all academic research, the authors have many people to thank for their help and support. Most of the field studies a century ago by Wanless are now in Badlands National Park. We thank Rachel Benton (retired) and Ellen Starck, chief geologists and paleontologists in the park, for the many years of their support. Ed Welsh, interpretive ranger, Alexi Richmond, seasonal preparator, and Bret Buskirk, intern, all at Badlands National Park has helped us in the field. Emmett Evanoff also thanks his chief photographer and wife, Kathleen Brill, and his long-time field assistant Justin Little for their help in relocating and documenting the Wanless photographic sites in the park. Vanessa Rhue and Chris Norris provided us with copies of Wanless' 1921 and 1922 field notes and photographs from the archives of the Yale Peabody Museum. Sammi Merrit at the University of Illinois Library Archives provided us with letters and an unpublished manuscript that has become Chap. 1 of this book. Alison Beninato guided us to approval of using skull reconstructions from the *Transaction of the American Philosophical Society*. John Murray of Murnor Studios in Coral Gables, Florida, aided in improving the contrast of some photographic plates from 1920 to 1922. Alice Levine and Melissa Lester provided helpful edits to the manuscript. We thank all these people for their support.

Coral Gables, FL, USA   Harold Rogers Wanless
Greeley, CO, USA   Emmett Evanoff

Contents

List of Figures

About the Authors

Harold Rogers Wanless is Professor in the Department of Geography and Sustainable Development in the College of Arts and Sciences at the University of Miami in Coral Gables, Florida. He earned his Bachelor of Arts in Geology at Princeton University in 1964, doing a Senior thesis on the petrology of the beach sands of Molokai Island, Hawaii. He completed his Master of Science degree in Marine Geology and Geophysics at the Rosenstiel School of Marine and Atmospheric Science, at the University of Miami in 1967, with an M.S. Thesis on the Holocene Sediments of Biscayne Bay, Florida. His Doctoral degree was in the Department of Earth and Planetary Sciences at the Johns Hopkins University in 1973, with a Ph.D. Dissertation on the Cambrian sediments of Grand Canyon, Arizona.He has been a faculty member at the University of Miami since 1971 and served a Chair of the Department of Geological Sciences from 1998 to 2017. Research by him and his students have focused on Modern to Pleistocene tropical sedimentation in Florida, the Bahamas, and the Turks and Caicos Islands; paleoenvironment reconstruction of Paleozoic marine sequences with a focus on Cambrian, Mississippian, Pennsylvanian, and Permian; and documenting limestone alteration (especially pressure dissolution) during burial and tectonic deformation. He served as principal advisor to students for 14 Senior theses, 18 Master's theses, and 12 Ph.D. dissertations. He has 64 jury referred scientific publications.Harold and his students have focused on refining past sea level dynamics and the role of catastrophic events (hurricanes) as these were critical to reconstructing our last 7,000 years of coastal and shallow marine history. He is using this understanding to better forecast future sea level rise because of global warming.

Emmett Evanoff is an Associate Professor in the Department of Earth and Atmospheric Sciences at the University of Northern Colorado, Greeley, Colorado. He earned his Bachelor of Sciences at the University of Wyoming in 1978, doing a Senior thesis on the stratigraphy of the lower Sundance Formation in the Bighorn Basin, Wyoming. He completed his Master of Science degree in Geology in 1983 at the University of Colorado in Boulder, with an M.S. thesis on the Pleistocene nonmarine mollusks and fluvial sediments near Meeker in northwest Colorado. He continued his studies at the University of Colorado, Boulder, working on a doctoral research project on the terrestrial gastropods, sedimentology, and stratigraphy of the White River Formation near Douglas, Wyoming. He earned his Ph.D. in 1990.Dr. Evanoff has been teaching at the University of Northern Colorado for 19 years. He teaches courses in historical geology, sedimentary geology, paleontology, and regional geology. He has been the advisor of four Senior theses, eight Master's theses, and on the committees of three Doctoral dissertations.Dr. Evanoff research has specialized in the stratigraphy and sedimentology of distal volcaniclastic sediments, including the Jurassic Morrison Formation of Colorado and Utah, the middle Eocene Bridger Formation of southwest Wyoming, and the upper Eocene/lower Oligocene White River sequence. He has worked in the White River rocks of Wyoming, Colorado, Nebraska, and South Dakota. For the past 27 years, he has worked on the stratigraphy of the White River Group in Badlands National Park, working closely with the paleontologists at the park. Having described and measured over 150 stratigraphic sections and mapped marker beds and lithologic contacts across the northern 40 miles of Badlands National Park, he is intimately familiar with the areas Sinclair and Wanless worked in the 1920s. He is currently working on a project locating and reimaging the photographic sites visited by Harold Rollin Wanless in 1920 through 1922.

Part I

A Summary of Fossil Hunting in the 1920s and an Introduction to Harold Rollin Wanless

Part I (Chaps. 1 and 2)

Abstract This section first is an insightful first-hand 1923 summary by Harold Rollin Wanless of his 1920–1922 field seasons in the White River Badlands of South Dakota. He provides not only the goals and achievements of the geologic and fossil collecting aspects but also a look at the nature of the landscape, homesteading families, Native Americans, climate, birds, wildlife, and plants.

This is followed by a look at the early upbringing and college years of Harold Rollin Wanless, seeking for a sense of those people and situations that aided in Harold's ability to subsequently excel as a scientist and person.

Fossil Hunting in the Badlands of South Dakota by Harold Rollin Wanless, 1923

Abstract

This is an insightful first-hand 1923 summary by Harold Rollin Wanless of his 1920–1922 field seasons in the White River Badlands of South Dakota. He provides not only the goals and achievements of the geologic and fossil collecting aspects but also a look at the nature of the landscape, homesteading families, Native Americans, climate, birds, wildlife, and plants.

Keywords

White River Badlands · South Dakota · Geologic methods · Fossil collecting methods · Badlands landscape · Climate · Birds · Wildlife · Plants · Native Americans · Homesteading families

Although the Dakota "Big Badlands" have been a treasure-house for scientific exploration for nearly 75 years, the general public has had little opportunity to visit them, and the term "badlands" does not seem to bring up a very distinct picture in the minds of most Americans – to begin with, there is nothing fundamentally "bad" about the "Big Badlands." This name is only a translation of "*mauvaises terres*," the term applied by the early French voyageurs to describe a very rough country, difficult to travel over and poorly suppled with good drinking water. The area with which this article deals is situated in the southwestern part of South Dakota, mainly in Jackson, Pennington, Washabaugh, and Washington counties, the two latter constituting a part of the Pine Ridge Indian Reservation. Most of the Badlands are located between the White and Cheyenne rivers which parallel each other for 20 or 30 miles. The "White River Badlands," as these are commonly called to distinguish them from other similar areas in Wyoming, Utah, Nebraska, and other parts of the West, were only to be reached by wagon road from the base of the Black Hills 40 miles to the west until 1907 when the Black Hills division of the Chicago, Milwaukee, and St. Paul Railway was opened, passing through the heart of the district. The Rapid City line of the Chicago and North Western Railway touches the northern border of the area at Wall, and the George Washington Highway, which is being rapidly made into a good graded dirt highway, follows the Chicago, Milwaukee, and St. Paul Railway. Thus, thousands of tourists [travelling] by automobile and train pass through this district on route to Yellowstone Park and the Black Hills annually. Several thriving towns, of which Scenic and Interior are the largest, have sprung up in the heart of the Badlands since the railroad was opened.

Strange as it may seem, in this area, where water and vegetation are scarce and few animals find a home, is preserved one of the finest records of an early stage in the evolution of our mammals. Here we find in hardened clay and sandstone the fossilized ancestors of the horse, camel, rhinoceros, dog, cat, and rabbit, as well as many groups which reached the climax of their abundance and diversity hundreds of thousands or millions of years ago. [They] then decreased both in number and variety, and finally disappeared from the earth. It is for the purpose of collecting the fossil remails of these animals and attempting to interpret the physiographic and climatic conditions under which the beds containing these fossils were laid down that the writer has spent the past three summers in this region.

Thaddeus Culbertson, a Princeton graduate, was [a] pioneer of Badlands scientific exploration. In 1850, a year or so after his graduation from college, ill health compelled him to find a change of climate. He secured employment from the American Fur Company and was commissioned by the Smithsonian Institution to make a collection of any interesting "petrifactions" he might find in the *mauvaises terres*. He reached Fort Pierre on the Missouri River, then the frontier post of the United States government, and set forth in a rickety wagon with two horses for the *mauvaises terres*, 160 miles to the west. After four long days of [traveling] he reached the basin of Bear Creek, near the present town of Scenic, and began a search for petrifactions. After a few hours he had gathered several "ugly brown objects which he was told were petrified turtles," and some other fossils, and as his animals seemed to be weakening and his supplies were growing low, he packed up his treasure and (and as his diary states) "as the area providing the petrifactions proved to be so circumscribed," he "hastened to finish collecting in one day" (see Culbertson and McDermott 1952). The weird topography of the Badlands fascinated him, and he engaged in speculation as to their origin. He was sure they were produced by fire and volcanoes, and repeatedly refers to the tremendous convulsions of the earth which must have taken place. He believed that great masses of the earth must have sunk to produce the broad badland basins which he saw so abundantly.

The area providing the "petrifactions" has proved to be far less circumscribed than Culbertson imagined, and collecting expeditions from leading universities and museums of the country visit the White River Badlands almost every year, so that now finely preserved fossil mammals from here may be found in most of the natural history museums of the country. John Bell Hatcher, the greatest and most successful collector of fossil vertebrates, visited the South Dakota Badlands almost every year for 10 or 15 years, and discovered an amazing amount of choice material. He is credited with the discovery of 150 skulls of *Titanotherium*, a giant horned beast resembling the rhinoceros in build and the mammoth elephant in size[1] (Fig. 1.1).

The badlands of the west consist mainly of horizontal beds of clay and silt which were formerly continuous, forming a high plain similar to that of western Kansas and Nebraska today. Remnants of this high-level plain are still to be seen on a large scale, constituting the "table lands" about 200 feet above the badland basins. These table lands are covered with 4 or 5 feet of sandy loam, which makes a good soil for crops. These table lands support most of the population of district. A very good yield might be obtained on the tables, except for the uncertainty of the rainfall and frequency of hailstorms. As an example, I may cite conditions on Hart Table, near Scenic, which has constituted a base camp for the Princeton parties during the past three summers. In 1922 no rain fell from the middle of May until the last week of June, then followed by the wettest July in recent years, with rainfall on 10 or 12 days. After this there was no more rain until the early part of November. The badland farmer thus faces the probability of seeing his crops nearly ruined by drought or hail about 3 years out of five. One sees a strange contrast in looking from the edge of a table land, with a fine field of corn or alfalfa growing clear to the rim, out over a desert-like vista of badlands where one might

---

[1]Titanotheres occur in the Chadron Formation, not the Brule Formation. Recent radiometric dating shows that they went extinct essentially at the Eocene/Oligocene boundary.

Fig. 1.1 Reconstruction of titanotheres by R. Bruce Horsfall. These were the largest beasts during the deposition of the White River Group, reaching a length of 13 feet and a height of 8 feet. They are a member of the odd-toed ungulates (perissodactlys) and are relatives of horses and rhinoceros. They occur only in the lower White River rocks of late Eocene age and are common in the *Titanotherium* beds now known as the Chadron Formation. (Originally published in Scott (1913, Fig. 160))

travel 30 miles without encountering a human habitation. The badlands themselves offer little in the way of farmland, through all the valleys of the more important branches of the creeks, or "draws," as they are called, have a fair covering of grass and sagebrush and offer good grazing for cattle if the season is not too dry. In the part of the district situated on the Pine Ridge Indian Reservation, most of the land is leased by the Indians to cattle owners with the stipulation that the number of cattle shall not exceed one animal to 20 acres.

The Badlands, as was stated above, have been, and are still being formed from the erosion of the high plain of which Hart Table, Sheep Mountain, and the other tables are remnants. They are formed purely by the destructive action of rain and running water. The normal rainstorm of the Great Plains is not a gentle shower, like most storms in the east, but a "cloudburst," generally lasting a very short time, and often accompanied by hail. As the drainage of the Badlands is almost perfect, every rivulet immediately fills with water, sending a small torrent to the draw to which it is a tributary. As each draw has thousands of tributaries, the volume of water in one [draw] will naturally be enormous. Often a 1- or 2-inch rainfall over fifty or a hundred square miles all finds its way down a single channel in the course of 2 or 3 hours. Because the transporting power of a stream varies as the sixth power of its velocity, it is evident that such a torrent must carry a great quantity of mud and silt from its bed and banks down to the White or Cheyenne rivers, and then on to the Missouri. Much of the color of the famous muddy Missouri is derived from the various badland areas which it drains. The fact that the banks of the draws are fine unconsolidated silt, clay, or sand, rather than rock, makes it much easier for the streams to erode their channels. Each depression or groove in the surface is accentuated by each rain and in the steeper slopes the water soon works out an underground channel connected [to] the surface by "swallow" holes, where surface clay has caved in to the underground channel, and where the material from the steep slopes is now washed into the underground streams. Many swallow holes are several feet in diameter, and they average from 5 to 10 feet in depth. As a stream gradually works up toward the dividing ridge it acquires more and more tributaries until eventually the area is perfectly drained. This is the youthful stage of the erosion cycle. The fact has been mentioned that the greatest development of badlands is situated where the White and Cheyenne rivers come closest together and run nearly parallel for a distance of 40 or 50 miles. It is natural that here, where the tributaries of two large rivers with a stream level about 400–500 feet below the level of the high table land are cutting rapidly toward each other, a very rugged topography should be soon developed, and high divides with very steep or vertical slopes are the natural consequence.

Another factor in the development of some of the more striking features of badlands topography is the presence of a number of beds of greater resistance to weathering than others. For example, there are several thin lenses of freshwater pond limestone which are much more resistant than the clay, resulting in ridges higher than the surrounding basin level or in benches along the sides of steep cliff faces. There are also found "fossil" stream channels, which are marked by a coarse hard sandstone, and may be traced for miles. These are so much harder than the clay that one sees in the course of a stream bed, which originally followed the lowest depression in the topography, outlined as a high skyline divide (Fig. 1.2). The upper third of the "clays" forming the White River Badlands consist largely not of clay, but of volcanic ash or dust (microscopic particles of pumice), quite thoroughly cemented so that where this series remains it form steep cliffs with sheer faces frequently over 100 feet [high], producing such castle-like topography as that seen in Sheep Mountain and the Wall of the Badlands (Figs. 1.3 and 1.4).

The writer has endeavored during the three summers in this field to reconstruct a picture of the district during the period when the sediments were being laid down, and the [ancient] animals were grazing on the plains. The first fact which impresses one in the study of the sediments is the great horizontality and continuity of the beds. Often a band 3 or 4 inches thick may be traced 10 or 15 miles. This means that the relief of the period was very gentle, comparable to that of the present high plains of western Kansas and Nebraska. Over this plain wandered several streams, in general leading away from the uplifted mountain areas to the west. As always, the high ranges gathered a greater amount of moisture than the surrounding plains and had a steadier and heavier rainfall. Because the relief was slight, it did not require an exceptionally heavy rainfall to cause one of the shallow prairie streams to overflow its banks and spread a thin sheet over a large section of the adjacent flat. The heavy pebbles from the mountains were carried in the channel where the current was strongest, and a thin sheet of mud was spread over a large area of the plain adjacent to the river. The channels shift their courses from time to time and the river was probably wide and shallow with several channels separated by sandbars, much like the Platte which at present flows through the high plain of western Nebraska. This stream has at many points a width of a mile or more, and depth of a few inches to a few feet. Scattered over the plain at irregular intervals were small ponds or depressions where water remained most of the time. In these various freshwater [ponds, shelled mollusks] and algae lived, forming thin lenses of limestone. These ponds doubtless resembled the prairie lakes which are now very characteristic of the eastern part of the Dakotas.

Most great periods of mountain building have been accompanied by much volcanic action and the Rockies were no exception to the rule. From volcanoes to the west, it is impossible to say just where great clouds of dust and "ash" (mainly fine particles of pumice) were thrown into the air and carried in the upper atmosphere for hundreds of miles. Finally, these showers of ash were brought down with the rain and spread quite uniformly over the plains. So important was this factor in the formation of the Badlands, that the upper third of the White River badland formations, the *Leptauchenia* beds, all contain from 50% to

Fig. 1.2 A sinuous narrow ridge capped by a coarse sandstone (dark gray blocky rocks). The sandstone is narrow and elongate and is called a sandstone ribbon. It represents the coarse channel deposits of an ancient stream. The channel sands were buried and cemented by calcite and silica. As a result, the sandstone is now more resistant to erosion than the surrounding mudstones. What was once the bottom of a sinuous stream has been exhumed forming a ridge. (Photograph taken 1921 in the headwaters of Corral Draw)

Fig. 1.3 Pinnacles on the south side of Sheep Mountain, the highest and most rugged of the uplands in the Badlands. It rises with nearly sheer walls about 500 feet above the surrounding basins. Its southern margin, shown here, has eroded resulting in many spires and pyramidal buttes and long thin ridges with deep canyons on either side. (Autochrome photograph taken in 1921)

Fig. 1.4 Pinnacles in the Cedar Pass area northeast of the town of Interior, with Walter Dew for scale. The rocks contain a large amount of volcanic ash which gives the rock resistance to erosion, resulting in near vertical cliffs and narrow pinnacles at the top of the Badlands Wall. (Photograph taken in 1921)

100% volcanic material, though there is no actual volcano known which could have erupted this material, nearer than 300 miles distant.

This is as near a picture of the condition of the Badlands and much of the eastern border of the Rockies during the Tertiary Period as can be drawn from the evidence in the field. Over this plain wandered great herds of mammals and giant land turtles. The assemblage of mammals probably resembles that of the "bush" of the high plateaus of Africa more than that of any other part of the modern world. It is hard to picture turtles as large and as numerous as they must have been at this time. The writer has seen the fossilized carapaces of six or more of them in an area less than hundred feet square and the total number encountered in the course of a season's prospecting could only be counted in the thousands. In length, the carapaces vary from about 6 inches to 3 feet. Next to the turtles in abundance was a small, hoofed animal called the *Oreodon*, which has left us no descendants (Fig. 1.5). It was probably allied to the ancestors of the deer, the sheep,

and the camel. Often the fragmentary remains of several hundred of these animals are seen in the course of the day. Figure 1.6 shows a skeleton of an *Oreodon* which was just being uncovered on a flat by the process of erosion.

The Princeton collecting parties of the past 3 years (of which the writer was a member) have reached the Badlands by means of a Ford car, because it is more economical, and because it proved to be very useful in the field. The first important problems for the collector are the location of the best fossil fields or "pockets," and the discovery of a water supply. Springs at the edges of the table lands, or shallow wells in the surface alluvium on the tables, supply most of the good water in the Badlands. Water gathers in depressions [on] the surface of the badlands after every heavy rain, but this is milky at best and after a few rainless days it becomes a white gelatinous-looking substance, which is about halfway between the liquid and solid state. True, it is wet, and the time comes with every collector when he would rather drink some of this solution of mud

Fig. 1.5 Reconstruction of two oreodonts by R. Bruce Horsfall. These were the most abundant and characteristic animals of the middle turtle-*Oreodon* beds. They were the size of a sheep and were even-toed ungulates (artiodactyls) related to camels, sheep, and pigs. Oreodonts were unique to North America, but the group became extinct before the Ice Age. During the time of deposition of the middle White River sediments (early Oligocene) the animals numbered in the millions. (Originally published in Scott (1913), Fig. 136)

than go without water altogether. Thus, water is a most serious problem, which must be solved before any actual field exploration can be done. In 1920 as all the Badlands were virgin territory for us; we were able to camp on one of the table lands near the town of Scenic, where a good spring supplied us with clear cool water and the town furnished us a handy base for supplies. In 1921 the territory near town had been quite thoroughly worked and we found it convenient to load up the Ford with about eighteen gallons of water and go off to a camp in the badlands for 3 or 4 days, returning when the water supply was exhausted. [During] the summer of 1922 it was necessary to go still further afield for virgin territory, so we had a ranchman drive us 20 miles into the badlands to the center of a rich fossil area and keep us supplied with water and provisions every 4 or 5 days for a month, while we made a good season's collection.

When one has arrived in the field and solved the water problem, the next question is "Where are the fossils to be found?" Ten minutes search almost anywhere in badland clays is likely to yield at least fragmentary bones or teeth of mammals, or broken plates of the shell of a turtle, but such random prospecting might yield nothing of collectible quality in a week. There are in the badland clay series, two or three horizons which apparently represent ancient land surfaces for a long period where the sun drew up groundwater to the surface and evaporated it, leaving the dissolved salts from the water as a cement for the clay. This [turned] the surface clay of that period into a hard rock shortly after it had been deposited, and it is in these "caliche" or nodular layers (Fig. 1.7), as they are called, that the fossil bones are best and most abundantly preserved. Prospecting for fossils generally consists in following one of these nodular layers where it outcrops along the sides of a high "table" or around isolated buttes. A bone-digger's paradise is found where a nodular layer forms the surface of the ground over an area of several acres. The [discov-

Fig. 1.6 Vertical view of the skeleton of *Oreodon culbertsoni* found in the valley of Bear Creek, August 2, 1920. This is just as the skeleton was found. The animal is apparently in a death pose with its neck bent back in an unnatural manner, and its hind limbs twisted over its vertebral column. Many skeletons occur in the Badlands but a few years of exposure to frost and rain will disintegrate a skeleton so badly that only few teeth and joints of limb bones remain. It is very rare to find a skeleton weathering out completely as this one. (This image [from the author's negatives] was published as Plate III, Fig. 1 in Wanless (1923), and is mounted as specimen at the Yale Peabody Museum, specimen VPPU.012565)

ery of] one or two such pockets may yield in two or 3 days a larger collection than weeks of prospecting narrow ledges of the nodular layer around the steep faces of badland buttes.

After a fossil specimen has been located it is necessary to get it removed from the clay or sandstone which forms its matrix, and to transport it safely to camp and then to the museum. In the removal of the specimen from its matrix, the greatest care must be exercised, for all specimens near the surface have been subject to the destructive action of freezing water for many years, and though they may look perfectly sound, may fall to pieces at the slightest touch. Therefore, it is necessary to protect each new surface of bone or tooth which is exposed with a thin coat of shellac, which hardens the soft bone. The shellac is generally applied on Japanese rice paper, a very thin paper, which serves as a strong binder to hold broken or cracked pieces together. In the case of a large or heavy specimen which is likely to suffer rough treatment before it has reached camp,

sections of burlap soaked in flour paste form a very effective binder, which can be easily removed after the specimen has reached its destination. An example of the necessity of a "paste cloth" to strengthen a specimen can be drawn from the case of a large and very perfect turtle which the 1921 party expected to collect. As the turtle was in a soft clay, the exposed surface was thoroughly shellacked with rice paper and allowed to dry, then turned over. The mass of soft clay filling the hollow interior of the shell was so badly [frost-shattered that] when lifted from the ground the turtle crumbled into a thousand pieces, and naturally was not collected. In a few fortunate cases it is possible to drive a car to within a short distance of a specimen, but generally the collectors [must] carry them the first few miles on their backs. This may become quite an ordeal, such as was experienced in carrying a 100 and 50 pound skull of a *Titanotherium* (Fig. 1.8) 300 feet up the steep side of Hart Table on a trail which was at sometimes so steep that it is difficult to

Fig. 1.7 A mudstone wall with numerous carbonate nodules and George Wiggan for scale. The nodules form from greater cementation within local areas of mudstones, typically along layers. The top layer is especially well cemented with calcite, supporting the ridge. Fossil bones can be abundant in these nodules. Many fossil skulls and turtles, where they occur in these nodules, are not compressed from compaction. This lack of compression suggests cementation of the sediment creating the nodules occurred not long after burial. (Photograph taken in 1921)

climb without a load. Or carrying most of the skeletons of two sabre-toothed tigers in over fifty separate packages for a distance of 7 or 8 miles into camp.

Now it is in order to present a picture of the life of the "bone-digger" in the field. Rising about five, breakfast is prepared on a portable kerosene "Optimus" stove which when not in use will fold up so as to go into a box only slightly larger than a cigar box. This type of stove has been extensively used by arctic explorers and generally gives satisfaction. A lunch is then put up and canteens filled with water, generally allowing each member of the party about two and a half quarts for the day in the field. A start is generally made about 7 o'clock, so as to reach the area which is to be prospected before the day becomes too unbearably hot. As a central camp is used for quite a long time, it is often necessary to walk from 5 to 8 miles before starting work. Prospecting generally consists in climbing up the side of one butte to look at a small outcrop of one of the nodular layers near its summit, then down a steep slope to the next draw, then up another and so on, until the noon siesta, or until a collectible fossil is found. Sometimes it is necessary to walk for long distances halfway up the steep slope of a butte, holding on when necessary, by means of a small hand pick which is always carried to help climbing.

The temperature generally rises to above 100 °[F] from about 10 to 3 o'clock, and when one is down in a narrow badland draw, where no breeze can enter and the mid-day sun is brilliantly reflected on the white badland clay, the collector may feel that he is experiencing a foretaste of the infernal regions. Culbertson's impression of the badlands seems very apt at such time: "Fancy yourself on the hottest day in summer, in the hottest spot of such a place, without water, without an animal, and scarcely an insect astir, without a single flower to speak pleasant things to the eye, and you will have some idea of the utter loneliness of the badlands." At lunch time the search

Fig. 1.8 George Wiggan excavating the jaw of a titanothere from the badland claystones in 1921. This fossil included a skull that was nearly complete and essentially uncrushed. The skull weighed 140 pounds, but it felt as if it weighed many tons when it was being carried to the top the steep slope of one of the flat-topped uplands 300 feet above where the skull was found. Two days were required to excavate the skull

for fossils is changed to a search for enough shade to allow one to rest with at least his head in the shade. We have at times walked a mile or more looking for a shadow from an overhanging ledge or dwarfed cottonwood tree large enough to protect us from the glare of the noon-day sun while we enjoyed our mid-day siesta. Trees are generally fairly common along the lower courses of the badland creeks, but the broad badland basins forming the sources of these streams, which are the areas generally explored for fossils, are often nearly or entirely devoid of trees. The shadows lengthen rapidly after 4 o'clock and the sunset and twilight are the most enjoyable part of the collector's day. The temperature is such that one feels no sensation of heat or cold. The heavens become afire with orange light as the sun slowly sinks behind the dark ranges of the Black Hills, reflecting its glory on the bare clay slopes of the badlands. As we watch the flame color gradually soften to a rosy pink and then fade to a violet and then the steel gray of night, the discomforts of heat and drought fall away, we stand awed by the glory and majesty of nature. Then after a strenuous day of physical labor, it is a pleasure to lie down on a solid bed under the starry canopy, unless perchance the frequent humming of the mosquito drives one to the protection of the tent.

The flora and fauna of the Badlands are quite specialized. The only trees which are at all abundant are the cottonwood, which commonly grows along the banks of the badland draws; the red cedar, which favors the steeper slopes, and the western yellow pine, which grows in thick groves on top of Sheep Mountain and some of the other high table lands. Many of the smaller plants are types characteristic of desert places. The most common are the sagebrush or *Artemisia*, the cactus, yucca, rabbit brush, and greasewood. The sagebrush is by far the commonest badland plant, [with] few slopes being too steep or dry for it [to grow]. In the early summer the Badlands present a large variety of attractive flowers which grow most abundantly in the alluvium around the edges of the high table lands. The blue foxglove, the mariposa or "gumbo" lily, a tall white larkspur, the thistle poppy, a variety of evening primrose, and the common milkweed are the most striking. Late in the summer the *Mentzelia*, a large annual shrub, blooms every evening. Its cream-colored, cluster-like flowers with a peculiar musklike fragrance attract a variety of moths and other insects. [See list compiled by Wanless of plants he observed in the Badlands at end of Chap. 16].

The commonest mammal is probably the prairie dog, although this is being rapidly exterminated by poisoning. A small gray chipmunk which can climb the vertical walls of badland buttes almost as fast as the eye can follow it, is also abundant. If camp happens to be located near a family of chipmunks the stores are likely to suffer, as we learned one morning when five chipmunks fled from our bread bag. The coyote, seldom seen but often heard, and the jackrabbit or Mormon antelope are other characteristic animals. Among the nocturnal varieties, the bobcat, porcupine, and badger are probably not rare, indications of their presence often being seen in the daytime. Until a few years ago, the Rocky Mountain [bighorn] sheep found the isolated pinnacles of the badland buttes attractive places to live, and Sheep Mountain, the highest and roughest of the tables, received its name from them. However, pursuit by hunters seems now to have entirely exterminated them in this district.

Among the birds the cliff swallow is one of the most abundant and characteristic. It lives in large colonies on vertical faces of cliffs or overhanging ledges of sandstone. Apparently new nests are built almost every year, as many colonies of several hundred nests each are found entirely unoccupied. Figure 1.9 shows such a colony near our 1922 camp site. It requires a sense which I am sure is absent in many human beings to dart with the speed of a swallow to a colony of nests, and without hesitation fly directly into the open hole of the right nest. The western meadowlark is the greatest optimist of the Badlands. Whether it is a scorching hot day without a breeze or a cloud in sight, or a cold rainy morning with no prospect of clearing, its loud, cheerful trill continually assures the "bone-diggers" that things will soon be better. The rock wren lives in little "swallow holes" and other small caves in the badlands and is also a great singer. The lark bunting, prairie horned lark, mountain bluebird, and magpie are other common badland birds. The burrowing owls are always to be seen around the prairie dog colonies and will generally bow politely two or three times and then fly away. These owls are reputed, with the rattlesnakes, to share the holes of the prairie dog, living on their young [burrowing owls eat mostly insects] [see listing of birds observed in 1920 at end of Chap. 16].

The prairie rattlesnake is the only poisonous snake, and it is fairly abundant, though one rarely encounters more than a dozen during a summer. They are not as large as the southern rattlers, rarely exceeding a yard in length, and they usually give plenty of warning of their presence. After shedding their skins, they are blind for a short time and then are likely to strike at anything which disturbs them, usually without warning. The bull snake is also common, and grows to be larger than the rattler, but is not dangerous.

The Pine Ridge Indian Reservation, occupied by the Oglala Sioux, contains a large part of the Badlands. The Indians live in small huts and occasionally make some attempt at cultivations, but generally depend on the government for their food and clothing. They are to be seen to the best advantage at the occasion of the annual Interior roundup, held in August in the heart of the Badlands. Generally, several hundred Sioux come from the reservation for this event, and camp all around the town. Most of them put on elaborately beaded and decorated garments for the occasion. Some of the women's dresses have rather interesting units of design (Fig. 1.10). I have seen one jacket almost entirely covered with nickels [five-cent coins] spaced in a geometrical pattern, while another used plum pits as the decorative unit, and still another used cockleburs. Porcupine quills and elks' teeth are the most popular and are both valued very highly. A jacket thickly studded with elks' teeth is a valuable article, as the teeth are only the upper canine teeth of the male elk, only two to an animal. Sometimes a hundred teeth are to be seen on one jacket. One can still see papooses carried by the Sioux mothers as one could in the days of our

Fig. 1.9 A colony of mud nests of cliff swallows. These nests are bunched together in colonies on vertical to overhanging cliffs of mudstone or typically on the underside of overhanging ledges of sandstone. Each hole opens downward, then there is a slight rise and a chamber behind the hole in which the birds nest. Nearly 150 nests are seen in this picture, which covers an area not over 36 square feet. (Photograph taken in 1921)

Fig. 1.10 Photographs of (**a**) Miss Running Horse and (**b**) another Sioux girl in traditional dress at the Interior roundup in August 1921. Their dresses are made of buckskin, quite elaborately beaded. Miss Running Horse's long necklace consists of bone pipes with some beads. Both have elaborately beaded capes and boots. (Photos taken by Rhoda Wanless in 1921)

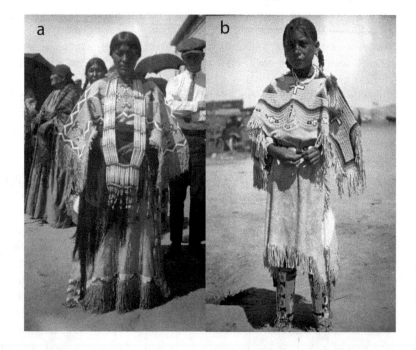

fathers, and the male elders often have their hair braided down to their waist. One may also see Sioux girls dressed in modern styles. The Omaha dance, which they perform in the central square of Interior to music supplied by a drum and a few singers, represents symbolically many old traditions of the Sioux, and to the bystander looks like our pictures of the war dances of the early days of the country. All the women and children stand in a big circle around the dancing men and keep perfect time with their feet. At the Interior roundup one also sees good riding exhibitions and other stunts, such as bulldogging of steers, calf roping, fancy roping, etc., in much more natural surroundings than those of the recent "rodeo" staged in Madison Square Gardens.

Most of the inhabitants of the Badlands are people who came west from Iowa, Minnesota, and the eastern part of the Dakotas, forming a late chapter in the westward advance of the Middle Border. For 30 years or more the Badlands have been a great stock-raising country and ranchmen with herds of thousands of cattle and sheep have till recently controlled most of the country. The upset financial condition of the country in the past few years, coupled with the effects of one or two very severe winters are tending to break the backs of the ranchmen. In the winter of 1919–1920, most of the ranchmen lost on an average nearly one-third of their stock. They are beginning to carry on farming on a small scale, and the table lands are becoming peopled with farmers, by whom cattle raising is carried on only as a side issue. Hog raising is also beginning to be popular and seems to yield much better and quicker returns than do the cattle. The high table lands have good soil, and when rainfall is sufficient, yield fine crops every year. Corn, wheat, oats, and alfalfa are extensively grown, and most farmers have small melon patches which give prolific yields. The badland valleys are not so satisfactory for agriculture and should be allowed to continue as grazing lands, for many cattle can be raised there on soil which is poorly fitted for any other purpose.

In spite of the farming which has been increasing lately, the cowboy is still the type of the country. Galloping into town on a fine western saddle pony, with bright colored bandana, burry chaps, and clinking spurs, the cowboy is indeed a picturesque figure (Fig. 1.11). Raised on the saddle, so that he can ride almost before he can walk, he is a splendid horseman. The vast openness of the plains seems reflected in this character [who] spends days in the saddle alone on the range. As people are not yet "too thick" in this country, the stranger is always welcome and receives all the hospitality that the humble shack or ranch house will provide. The cowboy has no room for the petty jealousies such as frequently creep into our more closely populated communities, and he takes for granted the fact that a man is to be trusted until he finds proof to the contrary. One of the most pleasant memories of the summers in the field are the evenings spent in a cowboy's ranch house listening to stories of the Indians, roundups, and other interesting incidents of life in the "great open spaces," of which Roosevelt and Harding were so fond.

The wonderful panorama one gets from a high butte or table land; mile after mile of treeless, canyon-cut, empty country; the thrill of discovery when a finely preserved fossil specimen is found; the cheerful call of the meadowlark; the

Fig. 1.11 A rider of the range. The cowboy, with wide-brimmed beaver hat, bright colored bandana handkerchief, large furry chaps, and clinking spurs mounted on a spirited but reliable cow pony presents a fine picture of a westerner. Shown is Archie, a hand on the Taylor ranch on Hart Table in 1921. (Autochrome photograph)

fragrance of the sagebrush in a narrow hot canyon; the pleasure from a drink of cool water when camp is reached after a day in the heat of the badland sun; the gorgeous succession of colors displayed in the sky as the sun drops slowly below the peaks of the Black Hills; wonderful nights on a solid bed under the myriads of stars which shine through the clear, cloudless sky.

These are the memories which do not fade.
FINIS

References

Culbertson TA, McDermott JF (eds) (1952) Journal of an Expedition to the Mauvaises Terres and the Upper Missouri in 2850. Smithson Inst Bureau Am Ethnol Bull 147:164

Scott WB (1913) A history of the land mammals of the Western Hemisphere. The Macmillan Company, New York, 786 p. Illustrated by R. Bruce Horsfall

Wanless HR (1923) The stratigraphy of the White River beds of South Dakota. Proceedings of the American Philosophical Society 62(4):190–269

Abstract

A summary by the current authors of the early upbringing and college years of Harold Rollin Wanless, seeking for a sense of those people and situations that aided in Harold's ability to subsequently excel as a scientist and person.

Keywords

Harold Rollin Wanless · Rhoda E. Wanless · Aurora · Illinois · Princeton University · Autochromes · William John Sinclair

Harold Rollin Wanless (1898–1970) was one of the most influential sedimentary geologists of the Twentieth Century and was known as an outstanding teacher and mentor, critical observer, prolific science writer, fine field geologist, and a man of uncompromising compassion and honesty. In 1920, the time of the diary that follows, Harold Rollin Wanless was a 21-year-old student at Princeton University (Fig. 2.1), just at the very beginning of his professional journey. His upbringing and his time at Princeton were critical steps in preparing him for his most successful career.

Harold was raised in Aurora and Chicago, Illinois, by his mother, Rhoda E. Wanless (Fig. 2.2), who was a naturalist and high school science teacher, specializing in nature studies. His father William Wanless, a lawyer, was killed in the 1906 San Francisco earthquake. Harold's mother had a very major influence on developing his intense curiosity, diligence, and communication skills in his growing up years. She instilled in him a love of nature, and especially the study and enjoyment of birds and plants. His early morning bird walks, binoculars in hand were a common part of time with his mother. Throughout his life, Harold sought to record the character and beauty of nature on film. Rhoda trained him well in photography with access to a variety of camera and film types, and he entered Princeton with a professional competence, even though he was still constrained by having to use individual 4½ by 6¾ inch format glass plates. Rhoda enjoyed handcoloring black and white photographic prints of the natural environment (see Fig. 17.5), but Harold eagerly took up using color glass photographic plates, called autochromes.[1]

Harold graduated from Waller High School in north Chicago. He finished high school at 15 and was held back a year by family before starting at

---

[1] An autochrome was a glass plate layered with three colors of pulverized starch (usually potato) as a color filter, charcoal, a panchromatic emulsion, and varnish. They became available about 1907 and were used until the late 1920s. Exposure times were slow, 1–10 seconds depending on lighting, so persons faces were easily blurred. https://www.scienceandmediamuseum.org.uk/objects-and-stories/history-colour-photography

H. R. Wanless, E. Evanoff, *The Diaries of a Bonedigger*,
https://doi.org/10.1007/978-3-031-25118-4_2

Fig. 2.1 Harold Rollin Wanless as a Princeton student with his typewriter on which he typed copious letters and the 1920 Badlands manuscript reproduced here. Most authors in the 1920s did not type their own manuscripts

Princeton University. During much of that year he stayed with relatives on a large apple orchard and at his mother's cabin, both in Michigan.

Princeton University provided Harold with an excellent education in geology. The Geology Department had nine faculty members and a robust curriculum.[2] Undergraduates were required to take physical and historical geology their first year and took what are still standard geology courses including structural geology, mineralogy, petrology, practical geology (a lab and field techniques course), optical mineralogy-petrography, and economic geology. Elective courses included two paleontology courses. Graduate courses included chemical geology (geochemistry), stratigraphy, vertebrate paleontology, invertebrate paleontology, and physical geography (geomorphology). The Geology Department had extensive research laboratories and a library containing over 6500 volumes. The department had an extensive teaching and research collection of rocks, minerals, and fossils in their museum. Field trips and field research projects were encouraged especially for seniors and graduate students. The professors that Harold mentioned in his letters while at Princeton include

doctors William B. Scott (vertebrate paleontologist and department chairman), Alexander H. Phillips (mineralogist), Charles H. Smyth (geochemist and petrologist), Gilbert Van Ingen (stratigrapher and invertebrate paleontologist), and Arthur F. Buddington (geochemist and petrologist).

Harold was a prolific writer, including many letters to a great variety of people, including his mother, some of which are included in the following chapters. Harold wrote very frequent and respectful letters to his mother and others through his time in college. In his sophomore year, he received a typewriter and, when he had it available, everything was typed (Fig. 2.1).

While at Princeton, he was a professional photographer for the *Daily Princetonian,* the student newspaper, and provided photographs for The Princeton Alumni Weekly. He was a staff member for the *Princeton Photographic Magazine* and the *Princeton Pictorial.* "Harold was often assigned to record some of the activities of distinguished visitors and thus met many very famous people" (White 1971, p 116).

During World War I, Harold served in the Princeton detachment of the Student Army Training Corps in 1918 and achieved the rank of sergeant.

Harold graduated as a geology major with high honors and his Bachelor of Science degree

---

[2]This information concerning the Princeton Geology Department is from the 1922 Catalogue of Princeton University.

Fig. 2.2 Rhoda Wanless in the Lake Michigan dunes near Muskegon, Michigan, summer of 1922

in 1920. He remained at Princeton, becoming a University Fellow in 1920–1921, Reader in Geology in 1921–1922, and Warfield Scholar in 1922–1923 (White 1971). He completed a master's thesis in the Spring of 1921, *Geology of the Rosendale Cement District, Ulster County, New York*, while working under the supervision of Professor van Ingen. This effort "is an example of his meticulous attention to detail and of his use of all these details in analyzing the geology of a region, in extracting general principles, and in determining geological history. The report consisted of two large volumes; the first of 279 typed pages includes detailed descriptions of scores of [stratigraphic] sections and deals in great detail with the stratigraphy, paleontology, mineralogy, economic geology, tectonic history, historical

geology, and paleogeography of the region. The illustrations are assembled in a second large volume of 122 superb photographic plates, many of them folding, mainly of strata and of tectonic and sedimentary structures, but some of topography and some of portraits of the Princeton staff [faculty] and students of historical importance" (White 1971).

Harold Wanless' advisor and mentor at Princeton was William John Sinclair (1877–1935). Sinclair was born and raised in San Francisco, California, and earned his doctorate from the University of California, Berkeley, in 1904. He worked with many of the paleontologists at the American Museum of Natural History between 1905 and 1914. He became an instructor of geology at Princeton in 1905. In 1920 he was an assistant professor of geology and started a 15-year research project on the fossil mammals of the White River Badlands of South Dakota. Princeton paleontologists had a long history of fossil collecting in the White River Badlands, starting with the collections of William Berryman Scott of 1882, 1890, and 1893, and those of John Bell Hatcher in 1893 and 1894 (Scott 1939; Dingus 2018). Sinclair was expanding the fossil collection in the departmental museum. He wanted to collect exhibit quality specimens for the museum and start a monographic study of the White River fauna in collaboration with Dr. Scott. In 1920 he was awarded a grant to travel to the White River Badlands of South Dakota. His field assistant was Harold Wanless, and they were accompanied by Sinclair's wife, Delia Sinclair, on their 1920 expedition (Fig. 2.3). Mrs. Sinclair often accompanied the two men into the field and cooked meals for the expedition.

Sinclair was an unusual paleontologist for his time in that he valued and made detailed geologic studies for interpreting the original paleoenvironmental settings for the fossils. By working with Sinclair, Harold greatly expanded his geologic and paleontologic abilities, eventually allowing him to excel in his field. Wanless expressed his gratitude for William Sinclair's mentoring in a letter to Mrs. Sinclair soon after Sinclair's death in 1935.[3]

"I feel much indebted to [Dr. Sinclair] for the opportunities he opened for me while I was a student at Princeton, for the encouragement and advice he was always so willing to give me. It was he who opened for me my first opportunity in the geologic profession, and I am not at all sure that I should have been able to take up this field of study and work if it had not been for his interest. He was also thorough in his training that I should observe things carefully and be sure of my conclusions before I announced [them]. I feel that I owe much to his thoughtful and painstaking efforts to train me."

Fig. 2.3 William and Delia Sinclair at the Hart Table camp in the White River Badlands, August 17, 1920

[3]Unpublished letter to Mrs. Sinclair, March 25, 1935, courtesy of the University of Illinois Library Archives.

Excerpt from the 1920 trip west by Sinclair (1920):

"Although explored seventy years ago by a Princeton man, Mr. T. A. Culbertson of the Class of 1847, and since then by parties from various colleges, universities and museums, it seemed probable that something new might yet be learned about the White River Oligocene formations exposed in extensive badlands in western South Dakota, southeast of the Black Hills, so a camp outfit was loaded on a Ford car and the writer and Mrs. Sinclair started westward on the eighth of June on the 1,827 mile drive to the fossil fields."

"This was a tame, though tiresome, procedure until the Missouri River was crossed (Fig. 2.4) and we discovered how very wet a dry country can be when cloudbursts are in season. On the river ferry, we met two parties bound for Yellowstone Park. On landing, each driver started westward by himself, but all halted at the first mud-hole, where the Buick party was hub deep in a plaster which cut like cheese and clung like glue. The (in)famous Pierre shale gumbo. After that, by mutual consent, everybody stuck together, literally and figuratively, camping together at night and by day hauling on the tow-line afoot, of milling in low through the slime with the gas-tanks and mud-guards of the heavy cars sweeping the ground, or pushing behind to dislodge a helpless car, stuck tight, with skid chains held firmly in the mud and wheels madly spinning."

2.1   Comments on the Diary

The trip to South Dakota was exciting for the young Wanless. While in South Dakota, he experienced the beauty and hardships that working in the White River Badlands provides in equal measure. Hundred-degree temperatures, violent thunderstorms, rattlesnakes, and the lack of water were balanced by the beauty of the Badlands, its flora and fauna, and the thrill of discovering the fossils. The discussions of the ancient fauna and geology with Sinclair were stimulating. Wanless would learn the techniques of field geology and paleontology that would allow him to complete his doctoral research project on the geology of the Badlands (Fig. 2.5).

He also got to know the inhabitants of the area: ranchers, shop keepers, homesteaders, even some of the Ogalala Lakota that made their way north from the Pine Ridge Indian Reservation. The Badlands had been settled for less than two decades by 1920, so many of the people he met were early pioneers and Indians trying to learn a new way of life (Hufstetler and Bedeau 2007). It was all very exciting for a 21-year-old geology student from the east. Fortunately for us, Harold

Fig. 2.4 Auto Ferry across Missouri River, Chamberlain, South Dakota with Ford Model T aboard

Fig. 2.5 Harold Rollin
Wanless in the White
River Badlands in 1921

kept a detailed dairy of his 1920 field experiences
and illustrated it with numerous black and white
and color autochrome photographs.

The field diary reproduced in this work was
compiled in the winter of 1920 and 1921 by
Harold Rollin Wanless. All photographs and
maps were taken by Wanless unless otherwise
noted. The field diary here is essentially the same
as Harold wrote it, but the current authors have
made a few modifications, such as correcting
grammatical and spelling errors, adding para-
graph breaks to some of the longer discussions,
and on rare instances, rewriting some of the more
lengthy and convoluted sentences. We have corrected the spelling of the fossil mammal genus-
names because Wanless spelled many of the
names phonetically as he heard them from
Sinclair. We have retained the names of the fossil
mammals that he used but are no longer used by
modern paleontologists. Harold discussed many
of the birds and plants that he encountered the
Badlands. We have annotated and corrected the
names of the plants following the South Dakota
plant guides by Johnson and Larson (1999,
2007). The letters that Harold wrote to his mother
and friends add much immediacy to the narration

and are not edited. Finally, the term Badlands
(capitalized) refers to the White River Badlands
of South Dakota. The term badlands (not capitalized) refers to the rough erosional landforms of
the area.

References

Dingus L (2018) King of the dinosaur hunters: the life of
  John Hatcher and the discoveries that shaped paleontol-
  ogy. Pegasus Books, 568 p, ISBN13: 9781681778655
Hufstetler M, Bedeau M (2007) South Dakota's rail-
  roads: a historic context. South Dakota State Historic
  Preservation Office, Pierre, 126 p
Johnson JR, Larson GE (1999) Grassland plants of South
  Dakota and the northern Great Plains (revised). South
  Dakota State University College of Agriculture and
  Biological Sciences B566, 288 p
Johnson JR, Larson GE (2007) Plants of the Black Hills
  and Bear Lodge Mountains, 2nd edn. South Dakota
  State University College of Agriculture and Biological
  Sciences B732, 608 p
Scott WB (1939) Some memories of a paleontologists.
  Princeton University Press, Princeton, 336 p
Sinclair, W. J. (1920) The Princeton geological expedi-
  tion of 1920 to South Dakota. The Princeton Alumni
  Weekly, 21 (12), p. 265–266
White GW (1971) Memorial to Harold Rollin Wanless
  (1898–1970). The Geological Society of America, 13 p

Part II

Journal of the Summer of 1920 on an Expedition of Princeton University for the Purpose of Collecting Oligocene Fossils from the Badlands of South Dakota

By
Harold Rollin Wanless
Princeton, March 1921

Part II (Chaps. 3, 4, 5, 6, 7, 8, 9, 10, 11, 12, 13, 14, 15, and 16)

Abstract   This is the previously unpublished journal/diary of the 1920 field season collecting Tertiary fossil skulls and other bones in the South Dakota Badlands, including illustrations from many of the black and white and color (autochrome) glass-plate images that he took. This interesting and detailed day-by-day description of the strenuous field activities is paired with insightful observations and evaluations of the landscape, the homesteaders, weather and climate, the methods and challenges in transportation and communication, and the birds, wildlife, and plants of the Badlands. Included are both the word-for-word journal and a number of personal letters to his mother and friends.

Abstract

Harold described the shifting scenery on the train trip west, in Rapid City, South Dakota, and back into the Badlands at Scenic. He begins his side commentary of the landscape, birds, animals, climate, and people that continues throughout. At Scenic, he offers his first impressions of the Badlands and hiking and fossil reconnaissancing in this stark country while waiting Professor Sinclair's arrival by automobile. Hand-drawn maps and photographs of Scenic shows a more thriving but just as delightfully barren town as today. Chapter ends with the first of many letters to his mother, providing very personal observations and evaluation of this naturalist.

Keywords

Train · Rapid City · South Dakota · Scenic · South Dakota

Sunday, June 20

I left the Northwestern Station, Chicago at 8 P.M., on the Chicago and Black Hills Express train. Gave up my lower berth to a lady with a sick husband on the way to Rochester, Minnesota for treatment, and was given an upper berth.

Monday, June 21

I travelled westward all day. We crossed the Mississippi River at Winona, Minnesota. There are high bluffs on both sides at a distance of a mile more or less from its banks. We went through Rochester, the great hospital city of the Mayo brothers, Mankato, a thriving town on the Minnesota river, Tracy, Minnesota, and Brookings, South Dakota, and at about 10 o'clock at night, Huron, South Dakota. Southern Minnesota is a great grain area, and in general very level. The extreme western part of the state and the eastern 30 or 40 miles of South Dakota are very marshy and seem to possess more lakes than dry land. The gradual thinning out of trees as one crosses Minnesota into Dakota is very noticeable. In eastern Dakota they seem to be mainly around ranch houses where they have no doubt been planted, and in stream valleys where they are natural. I slept at night in a lower berth.

Tuesday, June 22

I awoke at 4 a.m. when we crossed the Missouri River at Fort Pierre. We rode for miles up the valley of the Bad River. Cottonwoods were in the creek bottoms, and other trees seemed very scarce. We saw the first badlands near Wall to the south of the railroad (Fig. 3.1). We crossed the Cheyenne River, a recently swollen torrent, at Wasta. It had in the month before torn out bridges all along its course. We arrived at Rapid City at 9 o'clock, and I wired Mr. Sinclair at Scenic of my

H. R. Wanless, E. Evanoff, *The Diaries of a Bonedigger*,
https://doi.org/10.1007/978-3-031-25118-4_3

25

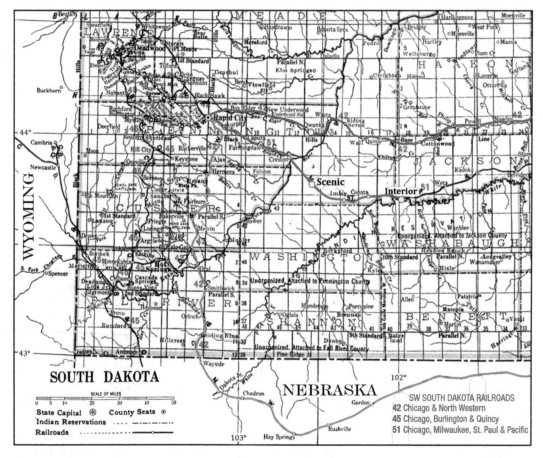

Fig. 3.1 Map of southwestern South Dakota in 1920. Railroads used by Wanless include his travel west to Rapid City on the Chicago and North Western railroad (blue line); his travel to Scenic and Interior on the Chicago, Milwaukee, St Paul & Pacific railroad (red line); and his return to Chicago on the Chicago & North Western railroad through Omaha (green line). (Map modified from the 1920 Hammond's Complete Map of South Dakota)

arrival. I hiked up to Hangman's Hill west of town. I was much impressed by the clearness of the air and the wonderful weather. There were many flowers and birds new to me.[1] The call of the western Meadowlark is very loud and attractive, more so than that of its eastern cousin. I saw a Magpie, Mountain Bluebird, Violet Green Swallow, and others. There is a fine view of the Black Hills from Hangman's Hill. Harney's Peak[2] and the Needles around it are seen to the southeast. The skyline to the east which is 50 or

more miles is entirely treeless, while to the west the hills are covered with quite a thick growth of evergreen trees, principally the yellow pine.[3] This gives a very dark color to the hills from a distance, and one can easily see the origin of their name. In the afternoon I met a prospective student at the South Dakota State School of Mines, which is at Rapid City. We walked over the Hangman's Hill ridge, and I took a few pictures with the Premo 3¼ × 4¼" camera[4] (Fig. 3.2). We found some petrified wood in a gully on the east-

---

[1] See lists at end of Chapter 16 of both birds and plants that Harold saw during the 1920 expedition.

[2] Now known as Black Elk Peak, the highest point in

South Dakota with an elevation of 7242 feet.

[3] Now known as ponderosa pine, *Pinus ponderosa*.

[4] Harold had an Eastman Kodak Premo No. 1 small format camera. It was his "pocket camera."

Fig. 3.2 (**a**) View north from [Dakota Ridge] near Hangman's Hill, west of Rapid City, showing the ridge in the distance, which with Hangman's Hill forms the sides of the gorge of the Rapid Creek. The trees are mostly Yellow Pine. (**b**) View of the central portion of Rapid City from Hangman's Hill looking east. (Photographs taken 22 June 1920)

ern slope of this hill. It was in two of three sections a foot or more in length and showed its annual rings. I afterward learned that the formation of this hill is the Dakota Sandstone. I spent the night in the Harney Hotel (Lodging $1.25).

Wednesday, June 23

I left Rapid City at 7:30 a.m. for Scenic, on the Chicago, Milwaukee and St. Paul Railroad. I arrived there a little after 9 o'clock. En route we crossed the Cheyenne River again near Creston. Scenic is a peculiar town to an easterner but is probably a typical western town. It has stores on both sides of a block and a half of the main street (Figs. 3.3 and 3.4). There are no trees in the town, nor even a shrub. At the time of my arrival everything was wet and muddy due to recent rains. Many tourist cars of various states were in town. Mr. Sinclair[5] had not yet been heard from. I engaged rooms with Mr. Kennedy, the hotel keeper. I bought crackers and hiked to the badlands just east of town and south of the railroad. I found no fossils but became acquainted with badland topography. I saw much more badland area to the south and east. In the afternoon I hiked to a high butte 1½ miles south of town (Fig. 3.5). I climbed and descended a canyon on its western side. I saw many swallow holes caused by the rapid cutting of rain waters in the soft clay so that their channels are for some distance underground, coming out only where the surface of the stream becomes fairly level. The slopes of the butte were very steep, and it is difficult to ascend or descend. I saw no fossils. The butte was worked the next to last day of prospecting [later in the summer]. At night there was no word from Professor Sinclair. There were tales from all the motorists passing through of being stuck in mud holes and hauled out. I also heard tales of Fords deserted along the road because their frame had broken in two. The roads are in terrible shape. There was a rainstorm about sunset with some lightning. I had supper at Mrs. Taylor's restaurant. She is assisted by Miss Grace Malloy. I received two cards from Mr. Sinclair, one from Epworth, Iowa, and the other from Sioux Falls, South Dakota. I slept in the hotel. I did not fall in love with the badlands on the first day in them. On my walks I saw many Horned Larks, Meadowlarks, and Killdeers, and my first Lark Bunting.

Letter to Rhoda Wanless (Harold's mother)

Thursday, June 24

I rose rather early and had breakfast, then hiked 2 miles west along the railroad track in the valley of Spring Creek. I saw no fossils. There were

[5]William John Sinclair (b. 1877, d. 1935) was a renowned vertebrate paleontologist at Princeton University. As has been customary at Princeton, he is called Mr. or Professor rather than Dr. throughout this diary.

Dear Mother:

I am now in Scenic and have been for about 5 or 6 hours. It is a regular frontier town with about 6 or 8 stores and a dozen houses. There is one brick building but no trees within a mile of more. There are two general stores, two garages, a restaurant, a hotel, a bank, a gasoline supply tank, and a railroad station. There has been a lot of rain lately. Mr. Sinclair has not yet arrived, but I just asked a man who drove in if he had seen a Ford from New Jersey, and he said he saw one 11 miles from here this morning. He ought to be arriving pretty soon if this is he. I have taken two strolls up into the badlands, the latter to a very high point. It is weird country, and I do not exactly love it yet. There is lots of short grass, cactus and some flowers, the most notable of which is the gumbo lily, a quite pretty flower. I have seen ten kinds of birds so far:

Red Shouldered Hawk, Sparrow Hawk, Mourning Dove, Killdeer, RedWinged Blackbird, Western Meadowlark, Horned Lark, Lark Sparrow, Lark Bunting, and Cliff Swallow. Of these, the Horned Lark is by far the commonest with the Killdeer probably next. The Lark Bunting is about the size of an Indigo Bunting or larger and is dead black in color with white wings. It flies up in the air and hovers fluttering while it sings. Its song is similar to that of the Song Sparrow. The Meadowlarks are nowhere as common as at Rapid City. I saw quite a lot of prairie dog colonies this side of the Cheyenne River.

Lots of cars have been stuck in the mud east of here and a few of them have been broken to pieces. They seem to have had about a week or more straight of rain and the roads are the worst I ever saw in my life. My duffle bag will probably come tomorrow morning. I got two cards from Mr. Sinclair, one from eastern Iowa written Friday and the other from Sioux Falls written Monday. I hope to see him tomorrow. I guess I will quit now and write more in a day of so.

With love, Harold.

numerous Shrikes and Meadowlarks. I returned to town and received my duffle bag, which had gotten a day behind me in transit. I hiked east of town about two and a half miles on the railroad tracks. I saw badland areas in all directions. I crossed a deep canyon of Jones Creek (a tributary of White River) and there found my first badland fossils, two or three large land turtles (*Stylemys nebrascensis*). Some were in quite good condition but weathered a rather dark brown on the surface. Some weighed over a hundred pounds. I returned to town over the low badland area of Bear Creek. I saw many attractive prairie flowers including the gumbo or mariposa lily (*Calochortus*), also some sage brush, cactus, etc. It became very hot by noon time. I waited at the hotel for two of three hours for news of the Sinclairs but received none. There were still many tales among the tourists of bad mud holes encountered. I saw many Indians in town. The Pine Ridge Reservations fence is only 6 miles south of town. Some of the men wore long braided hair hanging to their hips, and some of their wives carried their papooses on their backs as of old. Some of the women wore brightly colored blankets. They are of the Sioux Tribe.

In the late afternoon I started off for the gap[6] at the south end of Hart Table (name learned later), which is about 4 miles southwest of town. I found a small pond in which were a flock of 12 or more Wilson Phalaropes, beautiful shore birds. I also saw Arkansas Kingbirds and Brewer's Blackbirds. I reached the gap and ascended a canyon in it. Swallow holes were abundant in the canyon as they are in all badland canyons. Finally, I reached the top of the southeast side of the gap (Hart Mountain). There was a marvelous display of badlands before me as far as the eye could stretch. They were low rounded hills with apparently little vegetation (valley of Indian Creek and beyond). I found a few bone scraps on the way down, and some teeth just at the end of the canyon, in what we later termed the Spring Creek nodule field. I then returned to town. I took all three meals at Mrs. Taylor's. In the evening I took a ride with the Chevrolet agent from Rapid City in this car to Hart Table and back. The road is quite good. I slept with a young fellow from California because of a shortage of

[6]Now known as Saddlehorse Pass.

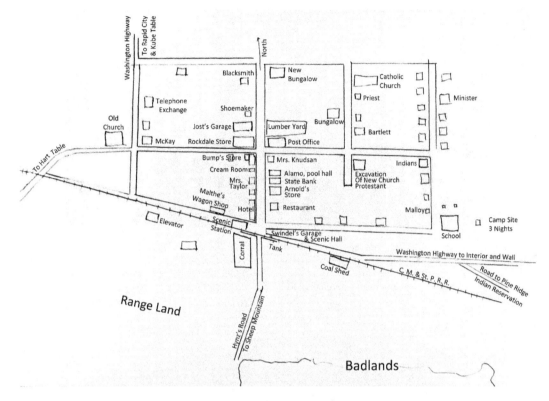

Fig. 3.3 Sketch Map of Scenic, Pennington Co., South Dakota, in 1920. Map relabeled for clarity

beds. I met an interesting young couple from California.

Friday, June 25

I ate breakfast with the California couple. The husband had come originally from Hightstown, New Jersey. There was still no sign of Mr. Sinclair. I fixed up a little lunch and took a milk bottle full of water and hiked east from town towards the badlands of the lower Bear Creek area. I found many turtles and bone scrap. In the morning Mr. Stanton,[7] head of the Stratigraphic section of the Geological Survey arrived with a young fellow from the School of Mines. Earl Taylor drove them down to Sage Creek where they found some quite good marine (Pierre Shale) fossils. I climbed the butte east of the valley of

Bear Creek (south of 71 Table) which was a very stiff climb. I also prospected about a mile of so along its base. It was a very hot day. I slept at night in the same room with the School of Mines student. There was still no word from Mr. Sinclair.

Saturday, June 26

It was a cloudy morning and cool and I did not intend to go out anywhere but after a while I started off south from town. I worked along and west of Reservation Road and found a great deal of scrap on the edge of what later turned out to be our rich Bear Creek pocket. I found much limb scrap and many turtles but no complete skulls. I worked on south through a big area of badlands to the east of Sheep Mountain.[8] I saw near Sheep

---

[7] Timothy William Stanton (b. 1860, d. 1953) was a premier researcher of Cretaceous invertebrate fossils in the early twentieth century. As Wanless noted, during 1920 Stanton was geologist-in-charge of the Paleontology and Stratigraphy Section of the U.S. Geological Survey. He would become the chief geologists of the U.S.G.S. in 1932.

[8] Now known as Sheep Mountain Table.

Fig. 3.4 Main Street, Scenic, from near the railroad track. Left to right: Mr. Kennedy's hotel, Mrs. Taylor's restaurant, Cream testing room, Mr. Bump's store, Rochdale store, Mr. Jost's garage; other side of street Mr. Bartlett's store and Post Office, Mrs. Knudsen's house, The Alamo, a pool room, State Bank of Scenic, Arnold and Wilhelm's store, restaurant, Mr. Swindler's garage, Scenic Hall. (Photograph taken 19 August 1920)

Fig. 3.5 Town of Scenic, seen from a low badland butte about half a mile southeast of town. A. Elevator. B. [Train] station. C. Hotel. D. Mrs. Taylor's restaurant. E. Mr. Bump's store. F. Rochdale Store. G. Scenic Hall and Swindler's Garage. H. Coal chute. I Catholic Church. J. School. K. Our tent and car. L. Kube Table. M. Main line of Chicago Milwaukee & St. Paul Railroad. (Photograph taken 19 August 1920). [Original ink drawn letters were fading and have been replaced]

Mountain, two coyotes, which ran at my approach. Sheep Mountain is very rugged and has many vertical-walled canyons of a height of a hundred feet or more. I climbed up the main canyon to a point at the foot of a vertical wall near the gap at the end of Sheep Mountain from which I could look over and see a big area of range land and badlands beyond. I was unable to reach the top, which consists of spires and pinnacles, nearly perpendicular. I left Sheep Mountain at 1:00 and arrived back as Scenic at about 4:30. The distance is about 10 miles.[9] There were many Turkey Buzzards[10] soaring above Sheep Mountain's spires and pinnacles and also some White Throated Swifts, which are beautiful birds with white tails and throats and very rapid flight.

It rained part of the way back to Scenic. The [south] end of Sheep Mountain is about 2 miles inside the Pine Ridge Indian Reservation and on the road back I passed three or four little buggies with Indian families. Some spoke quite pleasantly. When I arrived back at Scenic, I found a message from Mr. Sinclair to come to Interior. I learned that there would be an east-bound freight train the next morning that I could take.

On Wednesday there was a herd of over 1000 cattle being driven through on the range south of town to Wyoming for the summer. Lots of cowboys were in town with brilliant-colored bandanas, wide brimmed beaver hats, elaborate chaps, and clinking spurs. Also, there were 800 sheep south of town which were pastured there for 4 days waiting for cars to take them to Wyoming. Many had young lambs. They had been sold by Arnold, a big ranch on the reservation. I learned from much talk here that last winter was a terribly severe winter on cattle and sheep and that all the ranchmen of the section had lost very heavily. In the evening I took a ride with Earl Taylor (Fig. 3.6), Mrs. Taylor, and the young protestant preacher up to Kube Table[11] to pick up telephone wire. We then drove about halfway to Hart Table to repair the line where several poles had been knocked down by lightning in Wednesday evening's storm. I think the minister was younger than [me] and is a student at a theological seminary in Oberlin, Ohio. I saw a number of Burrowing Owls on fence posts along the road. This was my last night spent in a bed for almost 2 months. There is much talk of oil in this section as borings are being made at Hermosa (about 40 miles west), and it is thought possible that oil may be present here.

Scenic, June 26.

Dear Mother:

Well, after 4 days of waiting for Mr. Sinclair, I guess I am to meet him tomorrow at Interior. I have been putting in my time wandering around to various points near here. Today I made my most ambitious trip to Sheep Mountain about 10 or 11 miles from here. I am getting to like the Badlands very much. The weather has been very agreeable and there has been a little rain every day. It is very rough in the real badlands and Sheep Mountain is the roughest place of all. It has a number of spires and pinnacles almost 600 feet high looking like a cathedral.

I have come to know about half the inhabitants of Scenic and have made a good many friends among them. They are a very nice class of people. The main object of the town seems to be to supply tourists who pass through here. There has been more rain in this country lately than ever before and some cars are making only 30 of 40 miles a day. I have found quite a lot of fossils, mostly turtles and teeth. I have 5 or 6 kinds of teeth already and a number of small bones. The turtles are about 3 feet long and some quite perfect. Mr. Stanton of the geological survey was here yesterday working in the Cretaceous 20 miles east of here. He found some very fine Baculites shells about 2 feet in length. I understand that Mr. O'Hara[12] will not give nor sell copies of his bulletin as he has only two for himself. A student of his was here with Mr. Stanton and slept in the same room with me, so I heard a good deal of the School of Mines.

---

[9]This is the first of many 20-mile hiking days.

[10]These are now called turkey vultures.

[11]Named for Herman Kube an early homesteader.

[12]Cleophas C. O'Harra (b. 1866, d. 1935) was the first geologist and founder of the Geology Museum at the South Dakota School of Mines. He was appointed presi-

dent of the School of Mines in 1911. In 1910, O'Harra had published a bulletin titled "The Badlands formations of the Black Hills Region," which was the publication Wanless had wanted to obtain (O'Harra, 1910). However, in 1920, O'Harra published an update titled "The White River Badlands," which is still in press (O'Harra, 1920).

Fig. 3.6 Earl Taylor on horseback owned the ranch on Hart Table adjacent to where the expedition camped and made use of the spring water in 1920. (Autochrome photograph taken in 1921)

Near Sheep Mountain today, I saw 2 coyotes, but they ran away so fast, I could not see them long. They were as big as large dogs and light gray in color. Also 2 jack rabbits and a chipmunk. Have seen no snakes as yet. I have the following additions to my badlands bird list:

White Rumped Shrike, Nighthawk, Brown Thrasher, Arkansas Kingbird, Wilson Phalarope (?), Brewer's Blackbird (?), Mountain Blue Bird, White Throated Swift, Burrowing Owl, duck (?), Heron, probably Great Blue but possibly Sandhill Crane, Flycatcher unidentified, possibly Jay's Phoebe, duck unidentified possibly Blue Winged Teal, three hawks, each apparently different. The first probably a Red Shouldered or Red Tailed.

The second very light gray all over, possibly Ferruginous Rough Legged Hawk.[13] The third very dark all over and soaring over Sheep Mountain, possibly a Golden Eagle. Also 2 sparrows of a kind I had never seen before. I think it would be well to get a western bird book with the two dollars, and one showing water birds also if possible, as I am so unfamiliar with them all. The Horned Lark is much the commonest of any species and even walks down the main street of town.

The other night there was a little bronco busting excitement right in front of the hotel. The bronco was not very wild, however. Thursday a herd of over 1000 cattle grazed through here on their way to Wyoming. The last three days 865 sheep

---

[13]More likely a harrier hawk. Ferruginous rough legged haws are large with a reddish-brown back and leggings, and a white breast.

and lambs have been grazing here awaiting shipment to Wyoming. The table lands near here seem to be fine for wheat raising. I had a ride tonight with a rancher about 4 miles out who stays in town every night with his wife who keeps the restaurant I get my meals at. They are fine people. We fixed a broken telephone pole wire about 2 miles out. With us in the car was the minister fellow about my age who is in divinity school at Oberlin, Ohio. He is a very agreeable fellow. Have also gotten acquainted with a good many tourists coming through here. Two days ago, a young couple and another young man stopped here on the way from California to Pierre, S. D. They had been going three weeks and had come through Yellowstone Park. The husband was from Hightstown, N. J., 10 miles from Princeton originally.

I have not seen a paper since the Minneapolis tribune of Monday, it now being Saturday might. I do not think papers to speak of come to this town.

Had a bad cold for 3 days and coughed most of the time, but it is well now.

Have not heard from you yet and doubt if I will get any mail for a couple of weeks, as I will probably be at Interior for some time. Will be very anxious to hear [from you] when I can. Can you send the Western and Water Bird guides? I would appreciate it. Do not write me at Interior as I do not know how long we will be there. Was three miles inside the Indian Reservation today.

Love, Harold.

References

O'Harra CC (1910) The Badlands formations of the Black Hills Region. South Dakota School Mines Bull 9:152

O'Harra CC (1920) The White River Badlands. South Dakota School Mines Bull 13:181

Freight Train to Interior, June 27–29

4

Abstract

They camped quite near the present National Park Visitors Center on the prairie near Cedar Pass. He provides thorough descriptions of exploring for vertebrate fossils and collecting methods as well as camp cooking, water, and rather basic camping arrangements.

Keywords

Interior · South Dakota · Professor Sinclair · Paleontological field methods · Skull preparation methods · Camping · Cooking

Sunday, June 27

I had quite a discussion with the sheep owner who was taking the sheep mentioned above to Wyoming (Thermopolis). He was interested in oil prospects and intended to invest.

I took the freight train at about 9 o'clock for Interior. It was a freight for Kadoka and did not stop at Interior but merely slowed down. I managed to jump off with duffle bag, handbag, and camera with the aid of the man in charge of freight. I saw two Yellow Headed Blackbirds on the road near the station. I found Mrs. Sinclair[1] in camp on the schoolyard which is about two blocks from the station. They had had great expe-

riences wallowing through mud holes from Chamberlain where they crossed the Missouri River over 100 miles [east of] Interior. Their trip was 1830 miles in length and required 18 days, on one of which no travelling was done. When I arrived, Mr. Sinclair had gone off on a prospecting hike to Cedar Pass. He returned later in the afternoon having located a camp site about 2 miles from town and a mile from Cedar Pass. We cleaned the carbon from the carburetor and had supper. My first night spent in a sleeping bag.

Interior is a little larger than Scenic, and is somewhat more of a progressive place, I believe. It seems somewhat more of a farming community than Scenic. It is situated about a mile north of the White River, which is [on] the north boundary of the Pine Ridge Indian Reservation (Fig. 4.1). Interior has now an annual roundup which this year takes place August 18-19-20, in which there are competitive exhibitions of bucking, steer roping, bull dogging and many other forms of wild west competition. White and Campbell, real estate men here, seem to be quite boomers. In the evening Mrs. Sinclair and I attended the services of the Protestant church here, conducted by a younger preacher, a graduate of the Princeton Theological seminary in 1917.

Monday, June 28

I awoke at 4:30 and got dressed about 5:00 (Quite a change from my Princeton habits!). We picked

---

[1] Delia C. Sinclair (b. 1877, d. 1968).

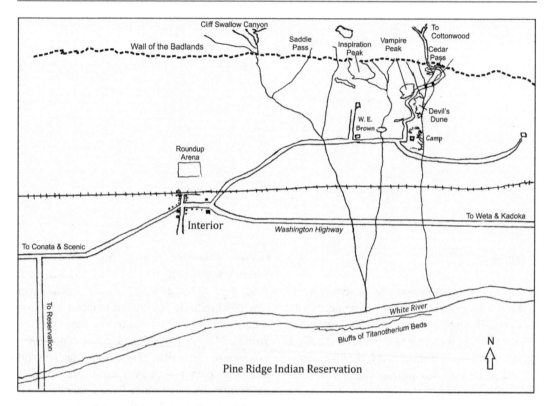

Fig. 4.1 Sketch map of vicinity of Interior, South Dakota, in 1920 [relettered for clarity]

up the camp equipment, took down tents and made two trips to the edge of the badlands about 2 miles northeast of town where we picked out a camp site. We pitched camp just at the edge of the low badlands about a mile south of Cedar Pass on the prairie (Fig. 4.2).

Many flowers were in bloom, notable among which were the mariposa lily, tall white larkspur, spiderwort, wild rose, snow on the mountain, small orange mallow, white evening primrose (Bone diggers sacred flower [according to] William J. Sinclair), and the yellow cactus[2] which is gorgeous in some places. Our equipment consists of two tents and a Ford car. One tent is a 7 × 7' box tent in which Mr. and Mrs. Sinclair live. The other, in which I sleep, is a shelter tent for the car. For cooking we have two kerosene oil stoves which can be taken apart and put in a small wooden box about 4″ × 10″ × 6″. These burn with a loud noise but give good heat. They are primed

with wood alcohol. Also, we have one oven. Mrs. Sinclair has a cot, Mr. Sinclair a bedroll, and I a sleeping bag. Breakfast we have fruit of some kind, oatmeal or cream of wheat, and bread or hot cakes. We have a large water can which holds ten gallons to keep us supplied while distant from water supply.

After camp was pitched, we had some lunch and then went out to prospect some of the neighboring buttes. We did not find scrap abundant but did succeed in finding one *Oreodon*[3] skull in fair shape. *Oreodon* is the type species of the middle

_____

[2]Prickly pear cactus, *Opuntia* spp.

[3]*Oreodon* was the genus name given by Joseph Leidy in 1851 to a small, now extinct artiodactyls (even toed ungulates) whose bones are incredibly abundant in the White River Group. However, the animal had already been named by Leidy in 1848 as the tongue-twister *Merycoidodon* and that name has priority. The name *Oreodon* was still widely used by paleontologist up until the middle of the Twentieth Century but was gradually replaced by the more general term oreodont, which refers to all the members of the Family Oreodontidae and not just *Merycoidodon*.

Fig. 4.2 Our camp south of Cedar Pass, near Interior, the morning before we broke camp. Note our proximity to the closest badland butte. Mr. Sinclair is seen working on the car. (Photograph taken 8 July 1920)

Oligocene beds. It was a small animal about the size of a small sheep (Fig. 1.5) and of a group of artiodactyls which has since ceased to exist. Little is known as to its habits . We prospected about half a mile of territory in the section east and north of camp but collected nothing. We quit about 6 o'clock. This section required a good deal of climbing as the best layer was near the top of the low buttes. The Cedar Pass section to the north of camp is a wonderful section as the Wall rises between 300 and 400 feet above the rest of the land (Fig. 4.3). The nights were beginning to be [moonlit]. The moonlight on the Badlands, illuminating the rough surfaces of the Wall[4] in the distance, gave a very weird effect.

Tuesday, June 29

This morning we prospected some of the section of the Badlands near camp, but a little north of the road. Mr. Sinclair and I each found a set of horse teeth which we collected. About ten in the morning on a small butte about 200 yards north of camp at the very top of the butte I found considerable material, and when Mr. Sinclair was

able to get there, it turned out to be a good horse skull with front teeth. We spent the rest of the morning and a part of the afternoon getting the skull out. He considers that [this] will make one of the best horse skulls in the Princeton collection when fixed up.

The method of extraction of a fossil is to loosen up as much of the surrounding matrix as can be loosened without danger to the specimen with a pick and then to use a digging tool (a tool rather resembling a small screwdriver) if the specimen is in clay, or a small steel chisel and hammer if the specimen is in rock. If bits of the specimen are broken off or loose, it is customary to soak that part of the specimen with shellac and then cover this with very thin rice paper. This, when it dries, very much strengthens the loose parts from the chance of loosening [even more during] carrying or in shipment. If the specimen is very large and heavy, it is generally advisable after the shellac has dried thoroughly to cover it with a paste cloth which is a flour sack or gunny sack (depending on the size and delicacy of the specimen) soaked in flour paste. This protects the specimen and if any pieces of it do jar loose, prevents their separation from [the specimen]. In most cases, the treatment after shellac has dried is merely wrapping the specimen in common cotton batten carefully, then in newspaper, and tying it up with a label. [The label records] the number

---

[4]The Wall is a line of badland ridges along the drainage divide between the White and Bad rivers. It is a formally named geographic feature. Wanless capitalized it in places and in others did not. For consistency it is capitalized in this discussion.

Fig. 4.3 Typical section of the Wall of the Badlands, situated just west of Cedar Pass, showing the very rugged character of the topography of the Wall, which continues along the divided between White River and Bad River, and further to the west between White River and Cheyenne River for about 50 miles, with only occasional passes. (Photograph taken 3 July 1920)

of the specimen, the species and part of the animal represented by the specimen, the locality where it was found, the date, and the name of the person who discovered the prospect. It is now ready for shipment.

The equipment which we carried with us consisted of a collecting sack, a canteen, and a pick except days when I wished to take pictures, when the 5 × 7 camera case, wooden folding tripod, and yellow color filter were added. The canteen held one quart of water (Mr. Sinclair's held three pints) and was carried on the left side of the shoulder strap. The collection bag, carried on our backs, contained the collecting equipment and lunches. The complete collecting equipment, although this was not always carried entirely in duplicate, consisted of newspaper for wrapping, cotton for wrapping, a shellac bottle, a shellac brush, a whisk broom, a curved digging tool, a small iron chisel, a small hammer, some rice paper for use with the shellac, and generally a Brunton compass, and when measurements were to be taken, a level.

This day we went back out to camp for lunch as we were working near camp. In the afternoon, we worked over an area to the northeast of camp, revisiting and collecting the *Oreodon* skull of the day before, as no others so good had been found. No new discoveries were made although considerable scrap was seen. The yucca flower was coming into bloom and is very beautiful (Fig. 4.4). The pick is very handy in climbing, as most of the buttes of this section, while not much over 40 feet in height, have very steep slopes. I find that the best way to prospect is to work along the base of these buttes, which is nearly a common level, and carefully look for bone scrap. When any is discovered, it is then best to look further up the slopes of the buttes, as of course all scrap weathers down the sides of the buttes . If possible, then I got to the top of the butte in question and found whether there was any trace of a prospect at the top or on the slopes, or whether it had "gone out," or been entirely weathered away. The horse skull mentioned was discovered in this way. It was "coming out" as the bone diggers expression puts it. The night was a beautiful moon night, giving a strange appearance to the surrounding badlands, and the section of Wall visible from camp.

Fig. 4.4 *Yucca glauca,* or the Indian Soap Plant, one of the beautiful and characteristic plants of the Badlands. A section of the Wall shows in the background. The Yucca blooms in July. (Autochrome photograph taken 3 July 1920)

Abstract

This begins the intense prospecting for skulls, horse teeth, and fossil turtles along the Wall area near Interior. Harold's photography reflects his awe at the spectacular peaks and spires in the Cedar Pass area and along the Wall. Mapping the sediments and sequences, their contained fossils, and their lateral correlation began here and later extended westward past Scenic. A small oreodont characterizes the upper, white, volcanic ash *Leptauchenia* beds. Figure 5.9 compares the rock units defined by Sinclair and Wanless with the currently defined rock units. Replacing brake pads on the Model-T, becoming accustomed to earth-shaking rattlesnakes, and enjoying the provision of milk and lettuce from locals help define working and living in the Badlands in 1920.

Keywords

Cedar Pass · The Wall · *Oreodon* ·
Leptauchenia · Rattlesnake · Model-T Ford

Wednesday, June 30

One of my duties from now on is to get milk before breakfast. Here we obtained milk from Mr. Brown[1] a farmer about a half to two thirds of a mile west of camp. He is a very pleasant man and has a wife and four children, Arbutus, aged 11, Margaret, aged 8, Walter, Jr., aged 6, and Buford, aged 3. Their house is a tar paper covered shack. He has a pump with good water, but we find it more handy to go into town to the pump by the schoolhouse and fill the water can every other day (weather permitting), as it is impossible to get close to Mr. Brown's pump with the car. Mr. Brown also offered us as much lettuce as we wished from his garden.

This morning we prospected a little closer to the Wall near the base of the butte known on post cards as "The Devil's Dune[2]" (Fig. 5.1). On a butte just south of this, I had my first contact with a rattlesnake. I did not know at first what it was as it was below the surface in a small shallow hole. It sounded like stepping on a small volcano, but it did not venture to attack me. It looked very large, but as it was coiled up, I did not get a good view of it. I did not kill it. We found a good deal of scrap including one collectible set of horse teeth. Early in the morning, I took a picture from the butte above camp of the section of lowlands in which we were working with the Devil's Dune, Vampire Peak, and the Wall showing in the dis-

---

[1] Walter E. Brown.

[2] "Devils Dune" is the high ridge just to the east of the visitor center and headquarters at Badlands National Park. The name is not currently used for this badlands ridge.

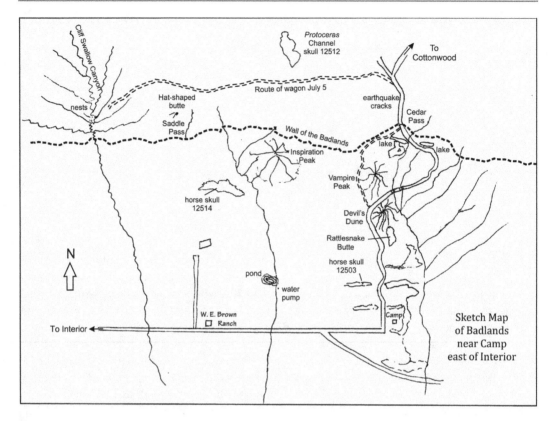

Fig. 5.1 Sketch map of Badlands near Camp East of Interior, Cedar Pass area

tance (Fig. 5.2). Before we went back to lunch, we went down between the Devil's Dune and Vampire Peak and got quite close to the Wall. It contains wonderful photographic material.

After lunch I took my camera with me and photographed a section of the Wall and Vampire Peak (Figs. 5.3 and 5.4). We then worked around a good deal of Inspiration Peak on which we spent the afternoon. We found very little material and nothing collectible. The layer which is exposed around the base of the butte in this section contains, in a lower belt, light greenish nodules and above them a belt of brown nodules. It is in [these brown nodules] where practically all the [fossil] material we have seen occurs. This is the "Red Layer" of collectors, of the lower *Oreodon* beds, middle Oligocene. The brown nodules are very hard and harsh to the touch and contain quite a bit of small root tracery. According to Mr. Sinclair, they much resemble the modern caliches being formed in northern South America and

probably owe their hardness to a leaching out of organic material and a replacement by limey material from below.

Late in the afternoon I took a short hike to the top of Cedar Pass alone. We went to town before supper. Another fine [moonlit] night. I changed plates for the first time under blankets, going headfirst into my sleeping bag. This makes quite a successful dark room.

Thursday, July 1

This morning I took my camera and went first to Inspiration Peak where I took an autochrome picture and another in black and white (Fig. 5.5). I had an accident with my camera here, as it was caught by a gust of wind, and rolled 10 feet or so down the side of the hill. It loosened several screws and lost three so that it required about an hour and a half to get it fixed. Finally, I got the pictures taken and then an autochrome and black and white picture of the east side of Vampire

Fig. 5.2 Early morning view from our camp near Interior, showing the low badland buttes near camp, and in the distance, the Wall of the Badlands. The high butte in the distance is Inspiration Peak. Note the alluvium dissection in the foreground. The arrow points to the place where one horse skull was procured. This would have been where the *Mesohippus bairdi* skull VPPU.012514 was found. (Photograph taken 30 June 1920)

Fig. 5.3 View of section of the Wall of the Badlands, just west of Cedar Pass. (Photograph taken 30 June 1920)

Peak and a picture of a section of the Wall (Fig. 5.6). I did a little prospecting before lunch and found quite a bit of scrap material, but nothing collectible. I went back to camp for lunch. In the afternoon, Mr. Sinclair and I worked off to the east about a mile or so south of the Wall. We did not find anything for some time, but later found the dentitions of three horses all within 25 feet of

Fig. 5.4 View of Vampire Peak from the west, from near the road to Cedar Pass. The upper nodular layer shows well in the middle of the picture. (Photograph taken 30 June 1920)

Fig. 5.5 Inspiration Peak (not named by me), a protruding section of the Wall of the Badlands, about half a mile west of Cedar Pass. Note extreme sparsity of vegetation. (Photograph taken 1 July 1920)

Fig. 5.6 A section of the Wall east of Cedar Pass, Interior, showing, in the foreground, the Pleistocene alluvium being dissected. This is a typical section of the Wall. (Photograph taken 1 July 1920)

each other. All of these were collectible. Scrap is in general quite scarce in this area and it shows signs of having been prospected in as some not very distant time. I changed plates at night again. We did not go into town. If gets very hot during the day. The water in the canteen tastes a little short of boiling by the hot hours of the afternoon (12 to 3 or 4). I wore dark glasses most of the first week or two as the light reflected by the white clays of the Badlands is dazzling.

Interior, South Dakota, July 1, 1920.

Dear Mother,

I have been trying to get a letter to you for a day or more, but we have been having to go to Interior for water almost every night. We get up at 5 in the morning or earlier. I could hardly do it the first day or so but am succeeding now. We have breakfast about six and I have to walk about a mile to get milk from a farmer. We are encamped near the edge of the badlands about two miles east of Interior and just a little over a mile south of Cedar Pass in the great Wall. Mr. Sinclair and I have been hunting fossils

for 4 days now and have so far collected 8 specimens worthwhile. There is lots of scrap but most of it too badly weathered.

Tuesday morning, I found a horse skull which Mr. Sinclair says is as good as any in the museum of this species – Mesohippus bairdi. We spent most of the morning excavating it. All our specimens collected so far have been horses. It has been very hot part of the time. Am burned and tanned.

We work from 7 a.m. to 6 p.m. with an hour off at noon. Have lots more to say but am writing in Sinclair's tent and they want to go to bed soon so must stop. Will write more before mailing if possible. Wish you could send me a couple pair of 10 cent store glasses as these seem a little broken. They are more satisfactory than the dark ones.

Love till later, Harold.

Friday, July 2
This morning Mrs. Sinclair accompanied us, and we started out for Cedar Pass and the wall. I took

a picture of the Devil's Dune and one of Vampire [Peak] on the way up (Figs. 5.7 and 5.8) and a beautiful subject up in Cedar Pass section of the Wall. Cedar Pass contains almost the only cedars in the vicinity and has a good many of them. Vegetation in general is quite different [here] from that in the low badlands. The pass is probably about 300 feet above White River valley and is the only pass in the Wall for about 5 miles in either direction. In many places, the Wall has almost perpendicular drops of 100 or more feet.

This morning we measured the thickness of the brown nodular clays of the lower *Oreodon* beds, the intermediate clays, the upper nodular layer, and the upper clays (Fig. 5.9). Of these, the upper clays with a thickness of somewhat over 100 feet were the thickest and the nodular layer with a thickness of 20 feet or less is the thinnest. Above the upper clays of the *Oreodon* beds occurs the *Leptauchenia* clays and the *Protoceras* sandstones (Figs. 5.9 and 5.10). The *Leptauchenia* beds are a very peculiar formation. They are very white in color and break up in small fragments rather resembling in appearance crushed Niagara limestone such as is used for macadam roads.

They weather with a vertical cleavage and tend to form vertical or at least very steep cliffs (Fig. 5.9). It is cut at all angles by a greenish sandstone[-filled] earthquake cracks, some being rather short in length and others being visible several hundred yards (Fig. 5.11). Some of these weather less quickly than the rest of the formation and form the cores of ridges. The type fossil of the white beds is *Leptauchenia*, a small animal of the *Oreodon* group (Fig. 5.12). Turtles are found rather commonly here. The fossils of this formation do not separate at all easily from the matrix because of great hardness of the latter. They are here easily visible from a distance because they weather a dark brick red in contrast to the white matrix. The Wall forms the divide between the drainage of Bad River to the north (joining the Missouri at Fort Pierre) and White River to the south. The section at the top of the Wall is one of the most barren sections I have ever seen. It is a dazzling white because of the color of the rock and the fact that there is not a tree for shelter nor hardly a blade of grass to rest the eye. There is much sagebrush and there are many small canyons to either side of the road. The White River

Fig. 5.7 The Devil's Dune, an isolated butte about a mile north of our camp at Interior, separated from the Wall, which shows on the right, by stream erosion. This shows, at the base, the Lower Nodular Layer, then the intermediate clays, the Upper nodular Layer, the upper clays, and is capped by the extreme Upper *Oreodon* beds, which seemed to be partly made up of volcanic ash. Professor Sinclair is standing on the side of the butte near the middle of the picture, giving an idea of the relative size. (Autochrome photograph)

Fig. 5.8 View of Vampire Peak from the south-east, showing much of the *Oreodon* series well. The road through Cedar Pass is seen just below the butte. Professor and Mrs. Sinclair are visible in the extreme lower right corner, inspecting a prospect

Fig. 5.9 Detailed image of the Wall showing the rock units defined by Sinclair and Wanless and the current rock units. Underlying photograph is the same as Fig. 4.3

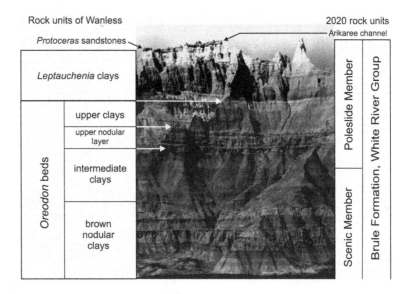

drainage that has cut into the wall has carved immense canyons as it is nearer streams than [is the] Bad River and consequently cuts more rapidly. We made no collections today. Returning from the edge of the Wall just above Cedar Pass, there was a wonderful view of the adjacent Badlands which formed the subject of a color plate and an Orthonon plate (Fig. 5.13). We returned to camp rather early and went to town before supper. I changed plates again at night.

Saturday, July 3

This morning was spent assisting Mr. Sinclair in the task of changing brake bands. This requires time and care as many parts must be removed before changing can be done. In the afternoon

Fig. 5.10 Section of *Protoceras* beds and *Leptauchenia* beds in the section at the top of the Wall of the Badlands, east of Cedar Pass. The *Leptauchenia* beds form the foreground, and most of the bluff in the background, and the [horizontal bedded] *Protoceras* beds the upper fourth of the bluff. The beds are volcanic ash material and a dazzling white color, and it will be noticed that not one scrap of vegetation shows in the picture. (Photograph taken 3 July 1920)

Fig. 5.11 A section of the top of the Wall, just above Cedar Pass, which is in the center of the picture. The foreground consists of the hard volcanic ash layers of the *Leptauchenia* beds, which are dazzling white in color, and of a quite hard character. Numerous earthquake cracks filled with greenish sandstone may be seen, running in various directions in the foreground. As shown, these have some influence on the topography. Through the Cedar Pass may be seen a section of the White River valley, and in the distance to the left, a part of Pine Ridge, which is 20 or 40 miles distant. To the left is a section of the *Leptauchenia* beds, capped by a ledge of *Protoceras* channel sandstone. (Photograph taken 2 July 1920)

Fig. 5.12 *Leptauchenia*, a small member of the *Oreodon* group, which is the type fossil of the *Leptauchenia* volcanic ash beds (Upper White River), exposed in Sheep Mountain and in the Wall of the Badlands. (Illustration by R. B. Horsfall, from Scott (1913), Fig. 201)

Fig. 5.13 View from the top of the wall of the Badlands at Cedar Pass into White River valley, showing the level nature of the valley, and the low badland fingers from the wall, extending out into it. The canyon of White River itself is seen in the dark band with trees about ¼ inch below the horizon line. The method by which badland canyons cut back into the divides separating them is beautifully shown by the three canyons in the immediate foreground. The Matterhorn like butte and the small lake caused by a landslide (of 1916) are seen in the very center of the picture. The Wall itself shows on the left [below which is the road]. (Autochrome photograph taken 2 July 1920)

Mr. Sinclair went prospecting northeast of camp and I went to the Cedar Pass again to take some more photographs. I took five photographs. [One of these photographs] was a beautiful badland cloud panorama [see the **Frontispiece**]. Yucca spikes are now in full bloom, and Fig. 4.4 is a picture of one of these in the Cedar Pass section. Neither of us made any collection this day.

Reference

Scott WB (1913) A history of the land mammals of the Western Hemisphere. The Macmillan Company, New York, 786 p. Illustrated by R. Bruce Horsfall

Abstract

Harold made a Fourth of July visit to the aptly named White River bordering the Pine Ridge Indian Reservation before a memorable holiday feast and fireworks with the Brown family in Interior. Up and west of Cedar Pass provided good fossil hunting especially for *Protoceras,* an ancestral horned relative of the pronghorn. While admiring a major cliff swallow nesting wall in a steep canyon, a torrential rain came up turning the slopes onto dangerous torrents for a time. Nodules of fossil skulls were emerging including full *Mesohippus* (3-toed horse) skulls. Important to the photography is mailing 6 color glass plates back east to be developed.

Keywords

Interior · South Dakota · Pine Ridge Indian Reservation · Brown family · *Protoceras* · *Mesohippus* · Cliff swallows · Cedar Pass

Sunday, July 4

The Brown's had invited us down to dinner today. In the morning, I went south to the White River about 2½ miles south of camp and beyond the railroad. This has a big valley but at the time I visited it, it did not have a great deal of water. It rises rapidly in flood time and frequently does much damage to bridges, etc., in its course. It has quite a quicksand floor and I sank to my knees at one point. There are numerous groves of cottonwoods in the creek bottom and other shrubs of less xerophytic character than those of the badland flora. I found numerous wild black currants which were quite sweet and tasted good in a country where wild fruit is so scarce. I also saw both the Arkansas and Eastern Kingbirds. The river is partly lined with bluffs, mainly on the south side, which consist of *Titanotherium* clays below the *Oreodon* beds (Fig. 6.1).

I returned to camp shortly before time to go the Browns (Fig. 6.2). The aunt and cousin of Mr. Brown[1] were also invited, as well as a boy about 17 years of age who works for [the] cousin of Mr. Brown. This other Brown family lives in a dugout about 5 miles north of the wall at Cedar Pass but has quite a large ranch, and I judge good crops and some stock. We had an excellent dinner with lots of good things such as ice cream (homemade), three kinds of cake, etc., which tasted fine after a week of camp diet. Western hospitality is certainly fine. Although the people generally do not have many of the comforts to which easterners are accustomed, they receive strangers with a

---

[1]The aunt and cousin were Alice and Charles Brown who had a homestead 2 miles north of the present Northeast Entrance to Badlands National Park. Their dugout homestead can still be visited at the Prairie Homestead, a site that is on the National Register of Historic Places (Crew and Heck 1996). The Brown's young worker was Bill Gannon.

Fig. 6.1 White River about 2 miles southeast of Interior. The bluffs on the south side of the river are in the *Titanotherium* beds. The river drains much of the badland area, and consequently its waters carry a great deal of clay in suspension, whence it gets its name. It here forms the divide between Jackson County and the Pine Ridge Indian Reservation, which is on the other bank. It was in fairly low water at the time the picture was taken but is subject to violent floods. There is a fairly rich zone of vegetation along its banks. (Autochrome photograph taken 4 July 1920)

warmth of friendship unknown in rural communities in the east. I walked to town with the boy who worked for Mr. Brown's cousin. A Mr. Ward,[2] professor of geology in the University of South Dakota, has been in town for 2 days or so intending to make something of a geological map of a section of the Wall near Cedar Pass and to start a collection of White River fossils for his University. After we returned to camp, I wrote a couple of letters. In the evening the children and Mr. and Mrs. Brown drove [over to our camp]

and I helped Arbutus, Margaret, and Walter shoot off firecrackers on the edge of the badlands just south of camp. They had lots of fun and are all very bright children.

Monday, July 5

We had planned to start for Scenic today, but as there was a little rain last night, and, since the roads were quite slick, we decided to put it off for a day. Mr. Brown rather wanted to locate the corners of some additional land on which he had filed north of the Wall, so he took us in his wagon up the road through Cedar Pass (Fig. 6.3). He also had in mind some big fossil bones which he had discovered. Little Walter, Jr., accompanied us. We left the road a mile north of the Pass and followed creek bottoms and low divides through a quite rough section to the west of here. Several creek crossings were steep and hard, but all were dry. The area of his land and where he had dis-

---

[2]Freeman Ward (b. 1879, d. 1943) was the State Geologist for South Dakota. He described and mapped the rocks in the Interior area in the summers of 1919, 1920, and 1921. Ward published his report in 1922 (Ward 1922), and his geologic map accurately shows that all the fossils Sinclair and Wanless collected around Interior were from the Brule Formation. The Chadron Formation (*Titanotherium* beds) does not occur in the area that they prospected for fossils.

Fig. 6.2 Brown family of Interior, probably taken 6 July 1921. Back row left to right Mr. Walter Brown, Rhoda Wanless, Alice Brown, Charles Brown, Mrs. Walter Brown; Front row, George Wiggan, Buford Brown, Margaret Brown, and Walter Brown Jr. Alice and Charles Brown were Walter's aunt and cousin who lived north of the Wall in the dugout house, [now a national historic site]

covered the fossil bones was about two to two and a half miles west of Cedar Pass, and a little west of Saddle Pass, an old saddle trail up the Wall, which has been now nearly obliterated. We left the horses near the edge of a magnificent canyon going down through the Wall into White River. We climbed a steep butte to the south of where we had left the buggy and found many teeth and large bone scraps in the immense blocks of *Protoceras* sandstone which had weathered down there. This is a channel sandstone and occurs at various points through the *Leptaucheria* clays. It is here very rich in fossil material, almost any small fragment of the rock containing a few bits of bone. Almost all the bone, however, is quite badly water worn and the fragments of bones and teeth are much scattered being in no association as to the animal to which they belonged. The type fossil of this formation is *Protoceras celer*, an ancestor of the modern prong-horned antelope (Fig. 6.4). This is a

coarse-grained green sandstone showing much cross bedding and clearly showing its channel structure in the cross sections produced by badland weathering. It is one of the most fossiliferous units of the Badlands.

While Mr. Sinclair and Mr. Brown were working on the location corners, little Walter and I prospected around the surfaces of the sides of the immense canyon. This canyon has almost vertical walls of 200 feet or more and because of the moisture from the rain the night before was bright green and red (Fig. 6.5). It was a veritable miniature Grand Canyon. At one point we found on the vertical cliff of the canyon a colony of 300 or more mud nests of Cliff Swallows, hence I gave this canyon the name Cliff Swallow Canyon for my own reference.[3] I found most of the dentition

[3] In 2003 Emmett Evanoff and his field assistant Terry Hiester hiked the length of this canyon for their first time. They noticed an abundance of cliff swallow nests in the

Fig. 6.3 Looking north from Cedar Pass along the road, showing the extremely rough and barren character of the topography accompanying the *Leptauchenia* bed outcrops. In the distance (left) a channel bed of *Protoceras* sandstone outcrops. Note that two or three clumps of sage brush are the only visible vegetation. (Photograph taken 3 July 1920)

Fig. 6.4 *Protoceras celer*, the type-fossil of the *Protoceras* channel sandstone (Upper White River), an ancestral horned antelope. (Illustration by R. Bruce Horsfall, in Scott (1913), Fig. 216)

of a *Mesohippus* and several other pieces of scrap in the clays of the Upper *Oreodon* beds here.

canyon and informally named it Swallow Canyon without knowing about the designation by Wanless 83 years earlier. We hope the name will become official.

About 1:30 or 2:00 a heavy storm came up suddenly and we all had no more time than to get back to the wagon before it was pouring rain. Mr. Brown hitched up and, with Walter covered by a big yellow slicker, started out. The rain was coming down in sheets, and in a minute or so Mr.

Fig. 6.5 Cliff Swallow Canyon seen from observation point where lunch was eaten July 6th. The top ledges to the right are the *Protoceras* sandstone, resting on the upper clays. The level of the top of the plateau is approximately that of the upper nodular layer, the vertical portion of the canyon is in the intermediate clays of the *Oreodon* beds, and the lower turtle *Oreodon* Layer. The canyon was colored a bright pink and green from the rain of the day before. (Autochrome photograph taken 6 July 1920)

Sinclair and I were absolutely soaked to the skin. We went on behind Mr. Brown's wagon for some distance till one of the deep creek cuts of the morning was encountered. The creek had changed from a dry bed to a rushing torrent about 2 or 3 feet deep. We had considerable difficulty in crossing as its bottom was slick and slippery, but we finally succeeded in effecting a crossing. We walked for a mile and a half in water between the tops of our shoes and our knees and met Mr. Brown at the road north of Cedar Pass. We rode a little with him but were too wet for comfort in riding, so we resumed walking, and walked the rest of the way to camp. After half an hour or so the sun came out and we were somewhat dried off when we reached camp. This, I understand, is a typical Badlands cloudburst. They are very local, as we heard later it did not rain at Scenic, 30 miles to the west. The rain makes the surface of the clay very slick, thus making walking difficult, but does not make a mud hole[4] as the clay is

practically impervious to water. After an hour of so it is usually pretty well dried off, and by half a day or so, the cracks characteristic of dry badland clays begin to appear.

Tuesday, July 6
This was a fine breezy day with a good many clouds. The night had dried off much of the rain of the previous day. Of course, the rain again prevented our starting for Scenic. Instead, we went back to the interesting section north of the Wall which we had visited the day before. I took camera and tripod and autochrome plates. We hiked up through Cedar Pass by an old road now deserted and I left Mr. Sinclair about a mile west of it to photograph a big channel of *Protoceras* sandstone in the *Leptauchenia* clays. This was a very remarkably diagrammatic channel. The appearance of the hill from the east was as shown in (Fig. 6.5). After this, I could not find Mr. Sinclair who was prospecting a section of the

---

[4]That is, the water does not soak all the way through the mud, wetting only the surface.

upper nodular layer south of here, so I went on to Cliff Swallow Canyon where I ate lunch and took an autochrome (Fig. 6.6) and a black and white picture. The subject was a wonderful one as it contained a section of 300 feet or more, and the coloring was rich. I also took a photograph of the Cliff Swallow colony on the vertical face of the cliff (Fig. 1.9). I prospected this area with some care, but without success and finally worked northwest to the headwaters of Cliff Swallow Canyon, almost a mile from the observation point.[5] Here I went down into the canyon, not very deep here, and worked down to its base. Some places it was so narrow I could hardly squeeze between its walls and at others it was filled with water, so I had to go almost knee deep, but it was a wonderful canyon and well worth going down (Fig. 6.7). I passed directly below the Cliff Swallow colonies and out to the open land south of the Wall (Fig. 6.8).[6] Here considerable

exposure of the *Titanotherium* beds below the *Oreodon* beds is found, but I found no *Titanotherium* material.

A little further on I found quite a fair exposure of the lower nodular layer and prospected some in it. Here I found a good skull in a nodule which might also contain a lower jaw. I thought it was *Oreodon* but took it back to camp anyway. It made a very heavy load with the nodule, camera and tripod and pick and collecting sack, all to be carried for about 2 miles. Mr. Sinclair at first pronounced it *Oreodon* and said it was not worth taking, but on later examination found that it was a fine skull of the three toed horse, probably *Mesohippus bairdi* (Fig. 6.9) and, on breaking part of the nodule off with a chisel, showed the complete lower jaw with front teeth. This would make it one of the best *Mesohippus* skulls in the Princeton collection.[7] Mr. Sinclair had also had the good fortune to find a male *Protoceras* skull

Fig. 6.6 Channel of *Protoceras* sandstone on the section back of the Wall between Cedar and Saddle Passes, showing a diagrammatic [cross] section of Oligocene stream channel now raised to almost the highest level in the vicin-ity. The blocks to the left are blocks of the channel sandstone which have been eroded down the side of the butte. A *Protoceras* skull was found in this vicinity. (Autochrome photograph taken 6 July 1920)

[5]This access point to the canyon is in the drainage near the modern Fossil Trail exhibit site at the head of Swallow Canyon. Please note, do not try to hike the drainage during or just after a major rainstorm.

[6]No titanotheres were found because the mouth of Swallow Canyon is actually in the lower *Oreodon* beds, not the upper *Titanotherium* beds.

in the area of the big channel described above. As

[7]The Yale Peabody Museum has both specimens, the *Mesohippus bairdi* is specimen VPPU.01254 and the *Protoceras celer* specimen is specimen VPPU.012512.

Fig. 6.7 Looking up Cliff Swallow Canyon from near its mouth, showing the nature of the bottom of a typical badland canyon. The vegetation showing is mostly sage brush. (Autochrome photograph taken 6 July 1920)

the Princeton collection had only female skulls, this was a fine addition. This was the best collecting day so far.

Interior, July 6, 1920

Dear Mother:

I am still here at Interior. We had intended to start for Scenic yesterday morning, but as there was quite a heavy rain Sunday night we stayed over a day. Then yesterday afternoon Mr. Sinclair and I and Mr. Brown, a neighbor farmer, and his six year old boy got caught in the heavi-

est rain I have ever been out in. The rain came down in sheets for about half an hour and every gully turned to a river several feet wide and 1 or 2 feet deep. We were absolutely soaked and, as we were over 3 miles from camp, had a good walk ahead of us. We were working near a new claim of Mr. Brown's up north of the great Wall. It is very wild country and there is a miniature Grand Canyon about 2 miles long and about 250 feet deep with vertical walls. I worked down it today and was able to follow it all the way. Mr. Sinclair found a quite good Protoceras skull this afternoon,

Fig. 6.8 View near the point where Cliff Swallow Canyon opens from the wall into the White River valley, looking downstream. All the badland buttes shown here are quite free from nodular layers, and has a very marked pink and buff banding. The stream shows quite wide meanders here. (Autochrome photograph taken 6 July 1920)

but I was not very successful. I lugged an Oreodon skull weighing about 12 or 15 pounds for about 2 miles, but it is not rare enough to keep. I also had my camera and tripod with me today. I have taken 5 color plates already and have them boxed up ready to send. Will probably mail it tomorrow. It is raining again now. Rains more in this county than even at Princeton. We do not know when we will get out of here. I have so far had only the one letter addressed to Interior, but we will probably be back in Scenic by the time you get this.

We had a treat Sunday in being invited to the Browns, out nearest neighbors, for Sunday dinner. They are very pleasant and though their house is just a small tar covered shack, they seem quite prosperous. They have had their holding here for 11 years. They have four children, Arbutus, aged 9, Margaret, aged 7, Walter, aged 6 and Buford, aged 3. The oldest two are considerable help to their mother and father, the oldest girl, Arbutus, rides a pony after the cows every night. They had an excellent dinner for us Sunday and also had relatives of theirs from over the Wall there.

I must close now, Wednesday p.m. as I am going to town and want to mail this.

With love, Harold

P.S. We will probably go to Scenic tomorrow.

Wednesday, July 7

This morning Mr. Sinclair and I started out to the place where I had found the horse but found no more specimens there. We then went up Saddle Pass to the top of the Wall east of Cliff Swallow Canyon. We found a great deal of *Protoceras* sandstone and much scrap in it, including some

Fig. 6.9 Horsfall's reconstruction of *Mesohippus,* the White River three-toed horse. (R. Bruce Horsfall illustration from Scott (1913), Fig. 152)

of *Protoceras* itself, but found nothing collectible. We spent a good part of the day around the area of the big *Protoceras* channel where he had found the *Protoceras* skull the day before but found nothing collectible. We examined other *Protoceras* exposures but also with no results. We finally collected a number of rock specimens to illustrate the various strata of the series and returned to camp quite early. We went into town this evening for the first time since Saturday. Mr. Sinclair obtained from Mr. Campbell, the real estate man, some fine calcite crystals from Rattlesnake Butte south of the White River. Mr. Brown had a bunch of fossil bones, etc., which he had picked up here and there, and [an] almost complete skull of a small rodent which Mr. Sinclair believed to belong to the genus *Steneofiber*,[8] which has not previously been listed from the Oligocene. If it is a new species Mr. Sinclair will name it *Steneofiber browni* in honor

of its finder. Mr. Sinclair had expected to start for Scenic Thursday morning, but as Mr. Brown wished more help with the section corners, he will go up above the Wall again with him. I mailed my first five autochrome plates to Gene Ericson today in a wooden box.

Interior S.D., July 8, 1920

Dear Gene:

 I have only time for a note as it is almost 8:30 and we go to bed by 9:00.

 I sent a box containing six color plates yesterday. They are in their box, wrapped in paper, packed in excelsior and in a wooden box, so they ought to come through all right. I have taken one of the remainders since but would like to hear from you before taking the others. Two of the six are old as they were purchased last November. One of these is from Princeton. Those two very likely will not be good, but the others ought to. I hope you can develop them soon

---

[8]Although still listed as *Stenofiber nebrascensis* in the Yale Peabody Museum collections, this specimen (VPPU.012513) is now known as *Palaeocastor nebrascensis*, an ancient burrowing beaver.

as I am anxious to hear of them. I have taken 27 others so far. It is great country, but there is too much rain, and it is very hot sometimes. We were caught in one of the worst thunder storms I have ever been in Monday afternoon over 3 miles from camp and had to walk in soaking wet. We expect to start out for Scenic, which will be our headquarters the rest of the summer. I have found two good horse skulls, each better of the species than the best in the Princeton Museum. We have not found so very much yet but have averaged about a specimen a day.

I would like to write more but it is bed time.

Best wishes, Harold R. Wanless

I am enclosing a folder of Interior Roundup which I fear I cannot see as we probably will not be here.

Thursday, July 8

This morning Mr. Sinclair went with Mr. Brown up over the Wall to help locate section corners. I hiked off east of camp about 2 or 3 miles and explored quite a large section of nodular layer. I found a good deal of scrap material and two or three broken or imperfect *Oreodon* skulls, but nothing collectible. It was an exceedingly hot day and not a very interesting section. This was our last full day at Interior. Mr. Sinclair made no collections this day.

References

Crew D, Heck D (1996) Prairie homestead, meet the Brons and their neighbors. Register-Lakota Printing, Chamberlain, 45 p

Scott WB (1913) A history of the land mammals of the Western Hemisphere. The Macmillan Company, New York, 786 p. Illustrated by R. Bruce Horsfall

Ward F (1922) The geology of a portion of the Badlands South Dakota. Geol Nat Hist Surv Bull 11, Series 22(6):80

Abstract

The 30-mile drive from Interior back to Scenic required repeated brake band adjustments. Permanent camp for the rest of the summer was on the edge of Hart Table (a flat-topped mesa in the Badlands) near the Taylor Ranch and a water spring. Collections near camp were challenging as most was broken scrap, with only a few good *Oreodon, Hyracodon* (carnivore), *Caenopus* (hornless rhinoceros), a small rodent, and fossil turtles.

Keywords

Scenic · South Dakota · Hart Table · Taylor Ranch · *Oreodon* · *Hyracodon* · *Caenopus* · Rodent · Fossil turtles

Friday, July 9

This morning quite early we broke camp. Mr. Sinclair took Mrs. Sinclair into town with a few things to check on the train. Then we loaded the rest of the camp equipment into the car, roped it in thoroughly and started. We had a very heavy load. The roads on both sides of Conata, the first town west of Interior, were very bad, and there were some deep ditches where culverts had gone out. It took us about 4 hours to make the whole trip of 30 miles or a little over,[1] as we had to stop frequently to tighten brake bands, and other minor details of the running of the car. We arrived safely a Scenic at about 3 o'clock and pitched camp near the schoolhouse there. Mrs. Sinclair deposited what fossils we had collected at Interior in Mr. Bump's store.[2] We got what information we could as to the possibility of getting with a car into Indian Creek and Corral Draw, etc. (Fig. 7.1).

---

[1] A drive of about 30 minutes today on South Dakota Highway 44.

[2] James Gay Bump (b. 1876, d. 1931) ran a store in Scenic and had a homestead on the north end of Kube Table. His second son, James D. Bump (b. 1903, d. 1959), was a geologist and paleontologist who in 1930 became the Director of the Geology Museum at the South Dakota School of Mines. In 1956 James Bump named the two subdivisions of the Brule Formation in the White River Badlands, the Scenic and Poleslide members. These two members are roughly equivalent to the turtle-*Oreodon* and *Leptauchenia-Protoceras* beds.

Interior, S.D. July 9, 1920

Dear Mother:

   We are about ready to start now. Mr. Sinclair had just started for the station with Mrs. Sinclair and I am waiting till he gets back to help pack the car for the trip to Scenic. The weather has been fine for the last two days so the roads should be in good shape again.

   Mr. Sinclair went up over the pass to help Mr. Brown locate his corners yesterday, and I prospected alone. Did not find much although I hiked a good many miles. The second day before, I picked up a big nodule with a skull showing on top which I thought was Oreodon. Oreodon is, in general, too common to take, but I carried it back two miles to camp anyway. Mr. Sinclair pronounced is Oreodon, and I thought all my labor was in vain, but the next morning he had thought it over and examined it again and found that it was the skull of a horse, Mesohippus bairdi, which is very good. He then wanted to knock off part of the nodule so as to lighten its weight (about 20 pounds) and in doing so exposed a complete lower jaw still in contact with the skull, which made it better than any other horse skull of this type on exhibition in the Princeton Museum. This is the second good horse skull I have found. The first had no lower jaw but had the complete front teeth of the upper jaw, which was also quite rare. Mr. Sinclair found a Protoceras skull which is the first made skull of this genus for the Princeton collection. We find lots of scraps of teeth and bones and turtles but in general do not take them, as they are not worth the expressage. Mr. Brown had found a small skull of a rodent which Mr. Sinclair believes to be a new species, and if it is, he says, it would be named Steneofiber browni, after the finder. If I do not look out, I will be having some species named after me.[3]

   I anticipate getting a lot of mail at Scenic as your one letter is all I have had in the last 19 days. As I wrote 19 cards [since] Rapid City, I hope to have some answers.

   I thought I told you that a student of Mr. O'Harra had said that Mr. O'Harra had only 2 copies of his bulletin left and that it would be practically impossible to get him to give one of them up. I believe he is getting a new edition out, so that there may be some available later. We expect to see him before the summer is over at Rapid City, so I will inquire then.

   I hope you have sent the orange glasses as the last piece of this pair is broken and I like them better than black glasses. Also, I hope you have sent the Western bird guides. I have seen the Rock Wren, Prairie Chicken, Dickcissel, Kingbird, and Say's Phoebe, I believe since I wrote you last. I also saw a sparrow yesterday which I did not recognize. The Rock Wrens are very common and chatter and sing all the time.

   I have sent my first color plates to Eugene and hope to hear about them soon. They are good subjects. Three days ago, I was working in a big canyon which had a colony of nests of about 200 or more Cliff Swallows. They are mud nests built on the face of a vertical canyon wall over 200 feet high. They are the only swallow here and one of the commonest birds. The White Shouldered Swift is also common around the higher peaks.

   I suppose you will be in Lakewood [their summer cottage in Michigan] soon if you are not already there. I envy you the trees there, as here it is often over a mile between trees.

   I must stop now as I see Mr. Sinclair's car coming back, and we will have to pack up the car for Scenic.

Love, Harold

P.S. What do you think of the democratic nominee[4]*? I heard who he was 3 days after nominated. Mr. Sinclair is just getting here now.*

---

[3]Prophetic words. The type specimen of the giant pig, *Archaeotherium, wanlessi* would be named for Harold (See the discussion of July 14 and in Chapter 17).

[4]The "nominee" refers to James M. Cox, the Governor of Ohio, who was chosen by the Democratic Party as its presidential candidate. Franklin Delano Roosevelt was his vice-presidential running mate. They lost to the

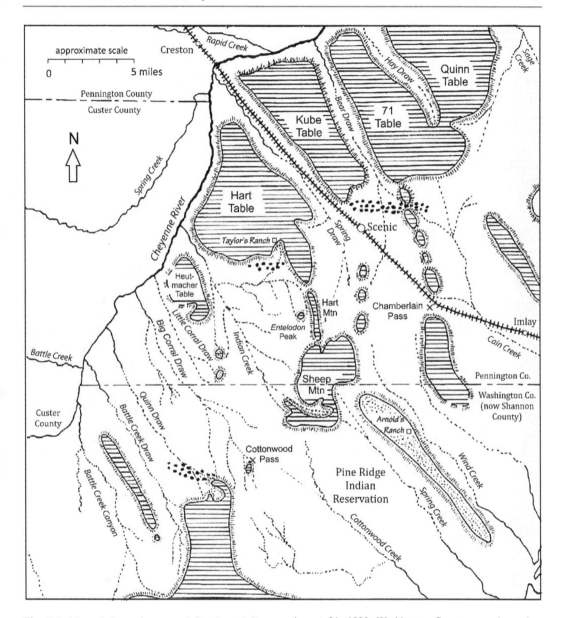

Fig. 7.1 Map of the region around Scenic and Sheep Mountain, South Dakota, showing the locations of geographic features described in the diary. This area was where Sinclair and Wanless worked from July 8 until August 21, 1920. Washington County was changed to Shannon County and most recently to Oglala Lakota County to reflect its association with the Oglala Soiux Indian Reservation. (Map is from 1922 expedition)

Saturday, July 10

Today Mr. Sinclair and I rode in the car to Hart Table and across it. The road is very good for South Dakota, much better than many sections of the Washington Highway.[5] The far side of Hart Table looks down into the valley of Indian Creek, which has cut a valley at least 200 feet below the Table. The base of it is here in the Pierre Shale, a marine black shale of the Cretaceous period, being the highest formation of the great

Republicans Warren G. Harding and Calvin Coolidge that November.

[5]The old name for the current South Dakota Highway 44.

Cretaceous inland sea represented in this area. We drove down the road going into the valley of Indian Creek and the road up it until we were stopped by a mud hole in the road. We stopped and for some time attempted to fill the hole with nodules from the Pierre Shale. Many of these nodules contained, we found, marine shells of large size, *Inoceramus, Baculites,* and others. We collected none of these. We decided to look further on the see what the road was like. It turned out that there were several bad mud holes ahead, so we decided it was impracticable to try to go further up Indian Creek in the car. We hiked up the road which has several creek crossings to Bowen's place about a mile up it. Mr. and Mrs. Bowen are both retired school teachers who have taken up farming. Have a very poor location in the creek bottom where any bad flood might wash their place out. Mr. Bowen offered to haul us through the mud hole, but we said we would hike further up first. We hiked about 2 miles further up Indian Creek to where it divides into three main parts, where we found another road. We decided it was impossible to drive up Indian Creek in the machine, so we walked back to the car and drove back up the hill to Hart Table. We drove to a point on the edge of Hart Table near Earl Taylor's place and sampled the water. He had a spring of good cold water which we liked. Then hiked down to the end of Hart Table, a distance of about three and a half miles. The edge of Hart Table looks down upon the biggest area of badlands in the United States or in the world probably. The valley of Indian Creek and its badlands form the foreground, and distant tables can be seen as far as the eye can stretch. To the west one has a fine view of the whole range of the Black Hills, crowned by Harney's Peak. Along the edge of Hart Table there were many gorgeous flower patches, chief among which is a large blue foxglove with brilliant sky blue flowers.[6] It is one of the most beautiful flowers I have ever seen and was not seen at Interior. The white larkspur, common milkweed and sunflower combine with this to make a striking color combination. We decided that a large area of badlands could be easily reached from a position at the edge of Hart Table, and consequently decided to pitch camp on the table just above Mr. Taylor's spring.[7] We then drove back to town and stayed again on the schoolhouse grounds at night.

Sunday, July 11

This morning rather early we again broke camp and in two loads moved to Hart Table to a position just above Mr. Taylor's spring. This makes a fine camp site. By noon we were established in our new site (which lasted for all the rest of our period of prospecting). It was a much more attractive site than that at Interior and far better than in the town of Scenic. I wrote a few letters and, in the afternoon, went down into the adjacent badlands for half a mile or so. The low badlands below camp are chiefly in the titanothere beds, and I found a great deal of *Titanotherium* scrap material. The *Titanotherium* was a giant horned beast something of the character and appearance of a rhinoceros (Fig. 1.1) but of a group which has since entirely disappeared. It was both the most abundant and largest of the animals in the lower White River beds. It probably attained a length of 13 feet or more and a height of 8 feet of so. It developed in the Eocene and assumed its dominant position in the lower Oligocene. Hatcher[8] has traced its evolution in three divisions of the lower White River beds by an increase in its size and a change in the character of its horns. At the top of the lower White

---

[6]The flower is the blue penstemon, *Penstomon glaber,* which has a similar flower to the foxglove.

[7]This arrangement was the beginning of a long association between the paleontologists of Princeton and the Taylor Ranch. Not only did Wanless and Sinclair return to this campsite, but it was also used by Glenn Jepsen, the successor to Sinclair, and by the Princeton graduate John Clark. Clark visited the Taylors as late as the 1970s.

---

[8]John Bell Hatcher (b. 1861, d. 1904) was the most renowned fossil collector of the late 1900's. He first collected in the White River Badlands in 1886 for O. C. Marsh of the Yale Peabody Museum. He later joined the faculty at Princeton University. He brought students into the Badlands in 1893 and 1894.

River beds, [its] scrap is probably more abundant than at any other point in the formation, the ground being in many places littered with vertebrae, femurs, tibias, ribs and other giant fossil bones. Above the division plane [contact] between the lower and middle Oligocene not a titanothere has ever been collected. No one knows where they went to or why they died, but it is considered quite possible that a change from moist conditions to extreme aridity may have accounted for their disappearance. The name Chadron Sandstone[9] has been applied to the *Titanotherium* beds and that of the Brule clays to the middle and upper White River Formation. I found that the lower *Titanotherium* beds, where they lie on the eroded section of the Pierre shales have apparently had a great secondary infiltration of iron compounds so that many bright colors have been imparted to them. This is especially noticeable in the section of the north branch of Indian Creek a little below camp. Colors of pink, red, blue, green, purple, and brown are all noticeable and make a very brilliant appearance for the rocks. There are also a great number of brown limonite scales throughout the lower part of the clays here which seem to indicate something of a secondary infiltration. The Pierre Shale contains many similar scales of gypsum, and all waters [emanating] from the Pierre Shale are so strongly alkaline as to be undrinkable. The rule in this country is that muddy waters are perfectly safe to drink, but clear waters, in general, are clear only because the alkali content has precipitated the mud. This was our first night spent on Hart Table.

Monday, July 12

Today Mr. Sinclair and I started down into the badlands below camp to prospect some of the *Titanotherium* beds (Fig. 7.2). We found a great

many bones of all sorts of the titanotheres, and in one place on a little hill scrap of *Hyaenodon*, sabre-toothed tiger, peccary, horse, *Oreodon*,[10] camel, and other animals but nothing collectible. We observed in the area much of the fossil excrement or coprolites of the Oligocene animals. It resembles modern dog excrement in appearance but is perfectly changed to rock. Mr. Sinclair says the most of this comes from carnivore animals due to the greater phosphate content of the intestines of these animals. We prospected almost all the morning in the *Titanotherium* beds with one of two excursions to the edge of Hart Table to look at the *Oreodon* beds. We found a great deal of *Titanotherium* material everywhere but nothing collectible. We ate lunch under some cedar trees in a side canyon from the main east branch of Indian Creek. We continued our titanothere prospecting in the afternoon but worked over to the edge of Hart Table and returned to camp quite early. We collected only a few teeth of *Ischyromys*, a small rodent. Mrs. Sinclair was quite sick tonight.

Tuesday, July 13

This morning Mrs. Sinclair was not feeling at all well and Mr. Sinclair decided to stay in camp with her. I started out at about 7:30. I worked down in the *Titanotherium* badlands below camp for an hour or two and then over along the edge of Hart Table for some distance. This is a hard section to work as there are canyons cutting the side of the table on the average about every hundred feet, which cause much difficulty and make progress very slow. About 11 o'clock I crossed over the low divide between Spring Creek and Indian Creek at the south end of Hart Table and worked in the upper nodular layer there for some time. I found most of the lower jaw of a *Caenopus* (a

---

[9]The Chadron Formation was named by N. H. Darton in 1899 for rocks in Nebraska and South Dakota that were essentially the same as the *Titanotherium* beds. He also named the overlying beds including the turtle-*Oreodon* and *Leptauchenia/Protoceras* beds the Brule Formation. Together these made up the White River Group.

[10]*Oreodon* was the genus name given by Joseph Leidy in 1851 to a small, now extinct artiodactyls (even toed ungulates) that were incredibly abundant in the White River Group. However, the animal had already been named by Leidy in 1848 as the tongue-twister *Merycoidodon* and that name has priority. The name *Oreodon* was still widely used by paleontologist up until the middle of the twentieth century but was gradually replaced by the more general term oreodont, which refers to all the members of the Family Oreodontidae and not just *Merycoidodon*.

Fig. 7.2 View of a small wide canyon in the valley of Indian Creek, showing at its head, an excellent development of skeletal sandstones in the *Titanotherium* beds; also examples of torrential cross bedding, in the right center; also how the head of this canyon is abruptly ended by a hard cap of sandstone. The vegetation of this canyon is quite typical of many badland side canyons. The grass as a rule rises only as high as the division between the Pierre shales and the *Titanotherium* beds. The upper sandstone of this picture is rich in bones of titanotheres. Sheep Mountain is seen at the left at a distance of about 5 or 6 miles. (Photograph taken 25 July 1920)

Fig. 7.3 Reconstruction of the hornless rhinoceros *Caenopus* by R. Bruce Horsfall. (From Scott (1913), Fig. 135)

large rhinoceros; Fig. 7.3) but nothing very good. About 1 o'clock I crossed over the southern continuation of Hart Table, Hart Mountain, which is somewhat higher than Hart Table and has much steeper sides. I went down from this into the valley of Indian Creek and worked up the valley of the east branch, which here runs almost parallel with and quite close to the edge of Hart Mountain. I found that an exposure of the brown nodular layer was here coming to about the ground level, and there seemed to be a great deal of material present. I collected no *Oreodon* skulls but saw ten of them in the course of about half a mile, besides the scrap of many others which had gone to pieces. There were hosts of fossil turtles, far more than I had seen before anywhere. I found and collected part of a *Hyracodon* skull with almost complete dentition, which we kept as the teeth were milk dentition. I worked for 2 hours very rapidly over the small canyons and exposure of this brown nodular layer around a high butte which came to a point at about the height of Hart Mountain (*Entelodon* Peak of future reference). I found almost the complete skeleton of an *Oreodon* with the vertebrae in order through the clay, but it was quite badly broken up. I took back one turtle which was perfect as a sample of what there are many more of. I returned by following the creek down about 3 miles past the cedars noted in the account of the preceding day, over a few badland buttes and up the road leading down from the Taylor's place to the Malloy's homestead, which is farthest up Indian Creek, back to camp. I arrived a little before seven. Mrs. Sinclair

was feeling much better and they had taken a little walk down into the Badlands not far from camp. Mr. Sinclair was much interested in the pocket I had discovered and said that we would by all means work it thoroughly if it proved as rich as it appeared. He said the Princeton collection contained no turtles as good as the one I had brought (from the White River beds).

References

Darton NH (1899) Preliminary report on the geology and water resources of Nebraska west of the one hundred and third meridian. US Geol Surv Annu Rep 19(4):719–784

Scott WB (1913) A history of the land mammals of the Western Hemisphere. The Macmillan Company, New York, 786 p. Illustrated by R. Bruce Horsfall

Finding the Giant Entelodont, July 14–17

8

Early in the prospecting in the *Entelodon* Peak area south of Hart Table, Wanless discovered an outstanding very large skull which turned out to be an entelodont, a giant pig-like animal. This was a new species that would be named *Archaeotherium wanlessi*. They also prepared and collected smaller entelodonts, several species of oreodonts, good *Hyaenodon* skulls (primitive carnivore), teeth of *Leptomeryx* (ancestral deer), *Palaeolagus* (ancestral rabbit), and *Colodon* (ancestral tapir). The big pig skull and others were recovered by a bone-jarring ride on a farmer's wagon.

Keywords

Entelodon Peak · *Archaeotherium wanlessi* · *Leptomeryx* · *Palaeolagus* · *Colodon* · Oreodonts · *Hyaenodon*

July 14 (Red letter day of the summer)

This morning Mr. Sinclair and I started down to the end of Hart Table and I conducted him to the area which I had found the day before. We soon found the *Oreodon* skeleton and worked over it some time. Mr. Sinclair decided to leave it as the skull was not very good and other limb bones were in quite bad shape. It was a clay specimen.[1] We found many more *Oreodon* skulls than I had seen in the area I had been over. We also found a large and a small turtle, each of which were practically perfect, which I had seen the day before. We decided temporarily that the large one was too big to carry the 5 or 6 miles back to camp. Just before lunch Mr. Sinclair picked up a fine *Hyaenodon* skull in a nodule washed out from its original position, but in almost perfect shape. The hyaenodons were representative of a primitive group of carnivores called creodonts, which died out in the middle Tertiary (Fig. 8.1). They varied from the size of a dog to that of a black bear (*Hyaenodon horridus*). They were probably about the most ferocious animals of Oligocene time. We ate lunch at the division of two branches of the creek, each heading up in the rough section at the north end of Sheep Mountain. After lunch Mr. Sinclair proceeded on up the eastern of these branches while I went back to look for an *Oreodon* skull I had discovered the day before which I thought might be a collectible specimen. Our criterion for the collection of *Oreodon* skulls is that they must have front teeth.

---

[1]A "clay specimen" is one which has been encased in swelling clays that over time swell when wet and shrink when dry, breaking the fossil apart.

© The Author(s), under exclusive license to Springer Nature Switzerland AG 2023
H. R. Wanless, E. Evanoff, *The Diaries of a Bonedigger*,
https://doi.org/10.1007/978-3-031-25118-4_8

Fig. 8.1 *Hyaenodon horridus*, a member of the primitive carnivores, Creodonta, which died out in the Oligocene, and *Leptomeryx*, a small deer-like artiodactyl of the White River. We collected five good skulls of various species of *Hyaenodon*. (Illustration by R. B. Horsfall, from Scott (1913), Fig. 277)

I had been gone scarcely 5 minutes when I saw a large nodule containing something shiny. I supposed it to be part of a large turtle but investigated. It turned out to be a large skull, and what showed were two over-lapping teeth showing that both jaws were present and closed. The teeth were weathered a bright blue. I thought it must be a skull of *Caenopus*, the large rhinoceros as I knew of no other animal as large. I was naturally quite excited and hurried to get Mr. Sinclair as soon as I could find him. He hurried back to the spot as quickly as possible and pronounced the skull one of an *Entelodon* or giant pig (Fig. 8.2), and said it was about the finest piece he had ever seen in the field in his years of collecting. We were both quite excited over it.

The skull was, I should estimate 24 inches long, including about the first three of the neck vertebrae (Fig. 8.3).

It was in a loose nodule with no other part of the skeleton showing anywhere. It seemed to be complete with the lower jaw with the exception of one canine tooth being broken, and an incisor missing. The weight of the nodule as we found it must have been somewhere in the neighborhood of 200 pounds. The animal is probably of the species *Entelodon* (*Archaeotherium*, or *Elotherium*) *crassus*. Princeton has the only mounted specimen of an entelodont in existence and that is of the *Archaeotherium ingens* from the *Titanotherium* beds collected in the Badlands by Hatcher. We decided to try to get one of the farmers to drive up the east branch of Indian Creek to this point as soon as possible to haul out the big skull as it was almost too heavy for two of us to lift from the ground, let alone carrying 5 miles to camp. We did not do much more prospecting that day but hiked back to camp by Hart Table. We felt that something well worthwhile was accomplished by now at any rate. We arranged with Mr. [David E.] Motter,[2] the farmer living about half a mile north of camp from whom we obtain milk, to go up Indian Creek either

---

[2]David E. Motter was a farmer on Harts Table in 1920 according to the Pennington County land-owner records. In his journal and letters, Wanless spelled his name both as Motter and Wotter.

Fig. 8.2 Horsfall's reconstruction of *Archaeotherium, Elotherium,* or *Entelodon ingens*, the giant pig of White River time, based on skeleton in the Princeton Museum. Note the long cheek phalanges, lower jaw processes, and also its root digging habits, as shown by the groves in the teeth of one of the animals in the Princeton collection. This will probably be renamed *Archaeotherium scotti*. [As a result of careful examination by Professor Sinclair, he renamed this species in 1921, in the Proceedings of the American Philosophical Society, v. 40, p. 468.] (From Scott (1913), Fig. 137)

Saturday or Sunday to get the big skull. I will after this apply the name *Entelodon* Peak to the high butte mentioned above, near the base of which the big skull was found.

Thursday, July 15

This morning Mr. Sinclair and I returned to the *Entelodon* Peak area and went over parts of the area prospected Wednesday again. We then worked north along the canyon at the base of *Entelodon* Peak. We found the skull of the smaller *Entelodon "mortoni"* about a quarter of a mile north of the big skull, and a little further on I found a small object which we considered probably a fossil bird's egg. It is about half the size of a hen's egg and somewhat crushed. We also found the teeth of *Leptomeryx,* the ancestral deer, and *Palaeolagus,* the ancestral rabbit. We discovered a good skull of the smaller *Oreodon, Oreodon gracilis,* and one or two of the larger *Oreodon culbertsoni.* We worked back on the north and west side of the *Entelodon* Peak ridge, following the exposure of the brown nodular layer. At about three in the afternoon, I found two or three teeth of the smaller *Entelodon* sticking up in the clay. We looked the prospect over and found that we could collect here an almost complete skull of the smaller *Entelodon*. We spent the rest of the afternoon excavating this, and Mr. Sinclair thoroughly shellacked the upper surface and sides. We left this overnight to dry to be finished and covered with a paste cloth the next day. We also found earlier a lower jaw of the larger *Entelodon clavus,* which we left but which Mr. Sinclair collected later in the summer. Mr. Sinclair said that this had already proved to be one of the richest pockets he had ever prospected.

Through the clays and nodular beds of the area of the big Badlands, there are a great many veins cutting in all directions filled with a very hard pink chalcedony, which is generally in the form of a double sheet with [in which are] crystals of yellow calcite. This breaks with sharp rough edges and walking on it is very hard on feet and shoes. As it is of course much harder than the

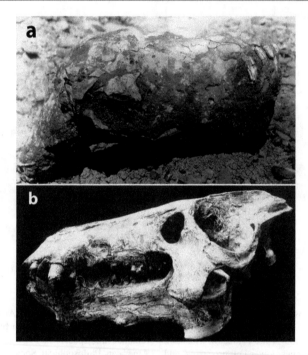

Fig. 8.3 (a) Skull (The type specimen of *Archaeotherium wanlessi* Sinclair, 1921) of big *Entelodon* taken where is was found near *Entelodon* Peak in this valley of Indian Creek. The top of the head, the front teeth, and the neck vertebrae are the only parts exposed. The teeth were a bright sky blue color. (Photograph taken 17 July 1920).

(b) Side view of the skull of the big entelodont when completely prepared, and as it is exhibited in the Guyot Museum. Note the long cheek [flange] and process of the lower jaw. [This is the type specimen of *Archaeotherium wanlessi* Sinclair, 1921. It is now in the Yale Peabody Museum as specimen VPPU.012522]

clays in whose cracks it occurs, it forms quite a blanket on the surface when eroded. The thickness of the veins is generally between one quarter and 2 inches. Little is known as to the cause of this great infiltration of silica. It was observed this day cutting through an *Oreodon* skull indicating that it probably took place sometime after fossilization of the bones here. The hollow marrow cavity of bones is frequently filled as is the inside (dentine cavity) of the teeth with material precisely similar in character. In places it forms large concretions and sometimes small rosettes. Many of the large concretions have inside a beautiful botryoidal structure. None of this chalcedony was observed at Interior. It appears to reach its greatest development in the section of the lower *Oreodon* beds but some veins of it occur at higher levels. It has also been found in some places quite abundantly with rounded edges, showing stress stream washing in the clays and

alluvial material which forms the (Pleistocene) surface of the plain.

Another interesting structure noted in the east branch of Indian Creek was a typical cross-bedded sandstone channel of *Titanotherium* age, going east and west in direction that was crossed at right angles by the modern channel of the east branch of Indian Creek. The beds of the middle *Titanotherium* clays in this section show much irregular banding of light pink and light green which are, however, not as pronounced as those in the canyon near camp as described above.

Friday, July 16
Today Mrs. Sinclair accompanied us to the point where our *Entelodon* skull was left to dry. Mr. Sinclair finished the excavation and I took a picture of him placing a paste cloth over it (Fig. 8.4).[3]

---

[3]Using this photograph and the photograph seen in Fig. 8.6, paleontologists from Badlands National Park

Fig. 8.4 Professor Sinclair putting a paste cloth on a skull of *Entelodon mortoni* (?) found on the west side of *Entelodon* Peak also showing the collecting bag and some of our daily equipment, and the method of preparation of a specimen after it has been excavated. It took several hours to excavate this specimen, as only a few teeth were showing. Using this photograph and another, paleontologists from Badlands National Park relocated this site in 2016 and found the tip of an entelodont tooth that seems to have come from the original skull. Note the nodules to the left. This was in a fine area of nodule exposures, and another *Entelodon* skull was found only about 100 yards distant from this one. Note the sparsity of vegetation, consisting of a few sparse clumps of sage brush. This is very characteristic of the Indian Creek badlands. (Photograph taken 16 July 1920)

This took most of the morning for him. I found another *Entelodon* skull less than 200 yards from this one, also at the base of *Entelodon* Peak on its west side. This was, however, not as perfect as the two best ones. We worked both sides of this canyon to the west of *Entelodon* Peak and found a great mass of turtles but nothing much collectible. Mr. Sinclair collected one tooth which he thought to be of *Colodon*, a very rare ancestral tapir of which a complete skull has never been found. On the west side of the canyon here our nodular layer was quite largely cut out by a channel of greenish sandstone (Fig. 8.5) which is probably equivalent to the *Metamynodon* channels of other parts of the Badlands. We started back for camp rather early as rain was threatening and arrived there a little before six. The trip tired Mrs. Sinclair quite a bit. We saw one rattlesnake in the morning.

Scenic, July 17, 1920.

Dear Mother.

I have a few minutes now before Mr. Motter, the neighbor farmer, comes with his wagon to take us up Indian Creek to where we have collected some big specimens to bring back. I think we will get into town tonight and I can mail this.

We have been working the Indian Creek section for six days now and last Tuesday I spotted an area which is the best I have yet come across. It is, however, 5 or 6 miles from camp and by the time we have hiked there with our pack of tools and pick and canteen and wandered around it all day and back at night, generally with 1 or more skulls in the bag, I am so tired I do not feel like doing anything but dropping down where I am and resting. Wednesday, I had the good fortune to make the best find of the

relocated this site in 2016 and found the tip of an entelodont tooth that seems to have come from the original skull.

Fig. 8.5 Channel sandstone in the valley of Indian Creek opposite the *Entelodon* Peak, showing the common toadstool-like formation resulting from small capping masses of sandstone. (Photograph taken 16 July 1920)

trip yet. It is a complete skull and lower jaw with a little of the neck of an Entelodon, a sort of giant pig as large as a modern cow. This is preserved in a nodule and shows 4 of the canine teeth, 2 upper and 2 lower locked together. They are big, blue teeth, shiny and almost 1½ inches long. The skull is about 18 inches in length. Mr. Sinclair said it is the best thing he has ever found in all his years collecting and the best preservation of this group in any collection. He says it assures the success of the summer expedition and would be worth $500. We have since found 3 more skulls and 2 lower jaws of the animal but none anywhere near as perfect. We have also collected 3 Oreodon skulls almost perfect having rejected 30 or more imperfect ones. Mr. Sinclair found a Hyaenodon (ancestral wolf) skull in almost perfect condition, and we have brought in 2 turtles in perfect preservation and have a bigger one to bring in the wagon this morning. We have also col-

lected jaw fragments of several other kinds. Mr. Sinclair is very well satisfied with our area and we will work here till prospects start to run out.

We are situated on Hart Table, about 250 feet above the level of Indian Creek valley which we overlook. This table is very grass covered and there are good farms of wheat, corn, oats, and alfalfa on the table. We are only about 100 yards from a spring on the place of Mr. Taylor, whom I think I mentioned on my previous stay at Scenic.

I will send my pressed flowers in a separated package. They are not all well pressed, but I think they will show a little of the flora of this region. The blue fox glove is the showiest flower here. Others of interest are white larkspur, yucca, cactus, the mariposa lily (now out of bloom), the thistle-like white poppy, spiderwort, and the garden flower of Chicago (called snow on the mountain). I believe I am including a bit of *cedar wood and a sprig or two of sage brush.*

Here we overlook the biggest area of badlands in the United States, stretching for 20 or more miles to the south and south-west. The whole Black hills dome shows on the west with Harney's Peak capping it, over 7000 feet high. In late afternoon light, we noticed that what appeared to be the northern Black Hills were really much further away. Mr. Sinclair says they are probably the northern Big Horn Mountains in northern Wyoming and southern Montana.

They were over 100 miles distant, so you see we have a broad outlook.

I have cut one rattlesnake half way through with my pick and seen 2 others which I did not attack.

Mr. Sinclair seems to think I can find things pretty well as I found most of that mentioned above except the Hyaenodon skull.

I guess I had better close now as Mr. Motter is starting to drive over in his wagon.

With love, Harold

Fig. 8.6 View of the nodule containing the big *Entelodon* skull just where is was discovered in the valley of Indian Creek near the base of *Entelodon* Peak, which shows in the background. The canine teeth show to the right of the nodule. These caused me to see the skull first. The exposure of nodules on the bank in the middle background is typical of this area. [The large skull-nodule is in the center foreground, to the right of the bush in the left foreground]. (Photograph taken 17 July 1920)

Saturday, July 17

This morning Mr. Motter drove us in his wagon down the Malloy road, and then, we picked the road up the east branch of Indian Creek for about 3 miles. In many places the creek cuts were steep, and it was hard to pick a passable way. As the wagon was springless, the jolting was very great, and it was much more comfortable to walk, which I did most of the way. We finally got the wagon right up to where the big skull was located and after I had taken two pictures of it (Fig. 8.6), loaded it in. We also procured the big turtle which I had discovered on Tuesday and loaded it in also. I then sat in back of the wagon on the hay to try to keep the big skull and other things from jolting too badly. We turned up the other canyon ("cow-bone canyon") to get the other two *Entelodon* skulls found there and loaded them in. We then started back for camp. The ride for the next 2 miles was the roughest ride I have ever taken. Between four skulls and a turtle to jolt me and a loose board to poke me in the back, I was not sure whether or not I would ever come through the ride. After about two or two and a half miles, Mr. Sinclair looked back and said he was afraid he would lose a bone digger if I did not get out soon, so I walked a good deal of the rest of the way. We trusted the skulls to behave themselves. We finally got back to camp about two o'clock in the afternoon and did no prospecting after that. Mr.

Sinclair did quite a bit of chipping with hammer and chisel on the nodule containing the big skull, exposing almost all of one side, and indicating that it was as perfect as it had appeared. We went into town before supper.

Reference

Scott WB (1913) A history of the land mammals of the Western Hemisphere. The Macmillan Company, New York, 786 p. Illustrated by R. Bruce Horsfall

Farmers, Fossils, and Heat, July 18–22

9

Abstract

Farmers, ranchers and families came to inspect the large fossil pig skull that was recovered. Though interrupted by threatening thunderstorms, they collected a sabre-toothed tiger skull and more, and were treated to spectacular sunsets. Terrible heat, reaching 115 °F in the shade, was unbearable in camp and resulted in catastrophic shrivelling and loss of crops.

Keywords

Heat · Farm crops · Sabre-toothed tiger

Sunday, July 18

This was our day off, and we did not go prospecting. In the morning I did quite a washing as I had not had a chance to before this. About ten we received quite a call from Miss Grace Malloy and her sister Leona. Grace has just graduated from high school and Leona has a year yet to go. They were much interested in our big skull and other things we had found. A little later Earl Taylor and Mr. Woods who work for him came and looked at the skull. It seems to be a matter of a good deal of local interest. Mr. Sinclair spent a good deal of the afternoon pounding off chips of rock, altogether relieving is of almost 50 pounds of surplus weight. In the evening Mr. Taylor who had been cultivating corn all day came and had a little supper with us and we enjoyed quite a visit with him. He is a fine true western type of man. His father came directly from England to South Dakota and settled at the "home ranch" on Rapid Creek about 25 miles from Scenic in the early days. And all the boys were brought up there. The Taylors have about a thousand head of cattle and consequently own or lease several sections of range land. They lost very heavily in last winter, a total of 235 cattle, I believe.

I wrote some letters in the afternoon.

Monday, July 19

This morning we worked on the west of the canyon west of *Entelodon* Peak for some distance, finding a great many immense turtles, some of them weighing 200 pounds or more, but found very little collectable. We did collect an *Oreodon* skull before lunch. In the afternoon we worked on to the south and west, near the northeast corner of Sheep Mountain, where the nodular layer runs under a grassy plain. We found one more *Oreodon* skull which was collectable and a good deal of scrap material, especially turtles.

It looked much like rain and became very cloudy so we decided we would rather be nearer Hart Table. We consequently went over to the gap at the south end of Hart Table, crossed the low divide to the valley of Spring Creek, and worked into the Spring Creek nodule field. This we found to be the greatest development of nodules we had met. It also seemed quite rich, but, as it was

Fig. 9.1 *Agriochoerus*, a long-tailed artiodactyl of the *Oreodon* group, which has some very peculiar structural characteristics. Note for instance, its claws. It is fairly common in the White River. (Illustration by R. B. Horsfall, from Scott (1913), Fig. 206)

threatening to rain quite hard, we did not stay very long. We did, however, stay long enough to collect a sabre-toothed tiger and the skull of an *Agriochoerus* (Fig. 9.1) an animal resembling and related to, the oreodons but somewhat larger. Thus, this day was also quite successful. We returned to Hart Table just above the nodule field and hastened to camp as rain was threatening, and it was raining quite hard west of the Cheyenne River. One feature of storms in this region is that you can see them coming miles away over the Black Hills and then see the rain coming down on the intervening portions and finally get nearer. The order of an average day is a brilliant sunny morning without a cloud in the sky, growing very warm by about 10 o'clock. About noon or 1 o'clock, a cloud (thunder-head) can generally be seen over the northern Black Hills or over Harney's Peak, which works up rapidly so that by 2:30 or 3:00, it generally obscures the sun and by four one sees a rainstorm approaching. However, the storm generally works off to the south over the Badlands or wears itself out before it reaches Hart Table. It is a frequent thing to see, at a distance, long streamers of rain going down through the sky but not reaching the ground. This day, however, we had a little 10 minute shower after which it brightened up, and we enjoyed one of the gorgeous sunsets for which the west is noted, the sun going down over the northern Hills, and the sky turning from golden to orange to crimson, lighting up all the little clouds in a blaze of color, which in turn gradually assumes tints of lavender and purple as the sun sinks further below the horizon. The sunsets are among the joys of the bone digger.[1]

Tuesday, July 20

This morning we returned to the area discovered Monday in the Spring Creek basin and worked it quite thoroughly. We found an immense field of nodules but were not fortunate enough to find anything collectable. We did, however, find a

---

[1] Source of the title of this manuscript. A "bone digger" is honorific to a vertebrate paleontologist.

good many scrap piles where animals had gone out.[2] As Mr. Sinclair says, nature is not a very good curator. After preserving these animal remains for a hundred thousand years or more it allows them to come to the surface and decay away till only a few small scraps of bone or teeth mark the fact that once a skeleton had lain there. As this area ran out in both directions, we crossed back to Indian Creek valley about noon to work along the side of Hart Mountain from the gap to connect up with our *Entelodon* Peak section. This was a very difficult section to work as there were big canyons cutting into the side of the mountain at frequent intervals and, as the exposures of the nodules were 60 or more feet above the base these canyons, it was necessary to do much climbing. In this section we found a good deal of scrap material but nothing collectible. When we had just about connected our section with our previous work on *Entelodon* Peak, I had the fortune to find two rather small *Hyaenodon* skulls within about 100 feet of each other. Each of these was in quite good shape. One is considerably smaller than the other, which also has a lower jaw present. Mr. Sinclair says this probably makes three different species of *Hyaenodon* skulls which we have collected. One or more of these may prove a new species. We again examined the location of the big *Entelodon* skulls for more of the skeleton but found none. We returned to camp a little early as it was very warm.

Wednesday, July 21

This morning Mr. Sinclair and I went to the area which we had left on Monday near the northwest base of Sheep Mountain and worked some distance of this section. However, the brown nodular layer was here under the surface and we found that the layers immediately above that, which weather rather greenish color are very barren in comparison. Other than collecting a sabre-toothed tiger skull without teeth which we had left on a previous occasion, we made no collec-

tion. On the way down, I killed a rattlesnake on the edge of Hart Table and collected his ten rattles. About noon the wind began to shift to the south and it became very hot and dry. We worked up to a place where the nodular layer was again exposed but found that here it was a solid sheet of caliche and did not separate as round nodules, thus the chance of finding anything in it was much diminished. Where we did find it in its normal phase, it was only in small exposures, as there were numerous large channels of greenish crossbedded sandstone, which frequently cut out the brown nodular layer. These channels, while they contained some scrap, were nowhere near as rich as the nodular layer. By 1 o'clock it had become so hot under the influence of a dry south wind that we could almost see the grass and sunflowers of the range shrivel under our feet. Mr. Sinclair had seen weather of this kind in Kansas and Nebraska and said that it generally practically ruined corn and the small grains if long continued. We returned to camp by about 3:30 as the heat was almost unbearable. We went into town before supper. This was one of the few hot nights of the summer.

Thursday, July 22

Today we stayed in camp as the heat was terrible. The hot south wind which seemed to dry up everything with which it came in contact, continued till quite late in the afternoon. It caused the skin and eyes to fairly burn from dryness. The dust became very bad around camp. This is very bad weather for crops, and wheat chaff was blowing into the milk [pail]. We heard later in town that thermometers in the shade registered 115°. As we had no shade, the temperature was probably several degrees higher. Late in the afternoon, the wind shifted to the southwest and by night it became again endurable. Clouds worked up from the Black Hills and things were practically normal. Mr. Sinclair finished cleaning the big skull and covered it with a paste cloth on gunny sack to protect it from damage.

I wrote two or three letters.

---

[2] That is to say, had weathered into piles of bone scrap.

Scenic, July 22, 1920

Dear Mother,

I received both your letter and package for both of which I am very thankful. I do not understand your hearing so little as I have written five letters and four post cards and so far, receiving only 4 letters. The dates and nuts and chocolate are a great treat, and Mr. and Mrs. Sinclair and I have enjoyed them greatly. I wrote the letter from Interior Friday morning and mailed it in time for it to get the eastbound train on Friday, but apparently you had not yet received it the following Wednesday morning. I received your letter written Tuesday night on Saturday night. We generally go to town on Wednesday and Saturday, but this week went on Tuesday also. I cannot tell anything about getting letters so just send them any time and I will get them when possible. Mr. and Mrs. Sinclair seem the think that blueberries would get smashed in the mail. I certainly wish we could get some of them, but perhaps it would be a waste of good berries. By the way, I noticed you sent the other box special delivery. There is no use in doing that as there is no delivery at Scenic anyway and had arrived the day before we went to town.

The last two or three days have been terribly hot, and, as a result, Mr. Sinclair and I are taking the day off today. Yesterday there was a hot wind from the south coming over the Badlands that felt like hot air out of a furnace. All the sunflowers in the Badlands were withering and drying up, and if there is no rain within a few days the corn and wheat and oats will start withering up and bring great loss to all the farmers around. The wind is somewhat cooler this morning but may become hot again before the day is over. I had a terrible night last night, as continual battle with mosquitoes from bedtime till morning light when the flies took the place of mosquitoes. This was probably because of the great heat.

The tent of Mr. and Mrs. Sinclair can be closed to shut out mosquitoes but mine cannot, as it adjoins the car, and they can get in anywhere under the car.

I believe, if you sent the overalls mentioned, they will help out as my trousers seem too hot and I am wearing my beer suit all the time. I have been carrying a bottle of water the last two days in addition to the canteen, but even at that I cannot seem to make it hold out long enough. The canteen water is almost boiling hot by 2 in the afternoon. We have gotten especially tired out the last few days as the area we have been working is 5 or 6 miles from camp and hiking this distance twice a day besides prospecting along the sides of cliffs for 5 or 6 hours a day with a pack and canteen on our backs and a pick with the temperature up around 100° or higher. We generally are parched thirsty when we get back to camp and ready to flop down whatever place is most convenient and not move till we have to. Since I wrote you, we have collected three Oreodon skulls, 2 saber-tooth tiger skulls (rather imperfect), one skull of an animal named Agriochoerus and 2 Hyaenodon, a primitive carnivore. He considers that we have been doing very well. Last Saturday morning, we went up to get the big skull of Entelodon and 4 other specimens too large to carry, in Mr. Motter's wagon. Most of the way there was no road and we followed the creek bottom. Mr. Sinclair and I walked most of the way up to help pick the road, but coming back, he had me sit in the back of the wagon to hold the specimens specially the big skull, from jiggling so much. I never have had such a bouncing in all my life. Every bone of my body seemed to rattle and crack, and I was not sure but what I would go under from the bouncing. It was a springless wagon over the roughest route I have ever travelled for 6 or 7 miles. However, we finally got back, and all was well again. Sunday, we did not *go out and I washed out my clothes, aired my bed and wrote [letters].*

Please tell me in what kind of shape you get the magazine with the pressed flowers.

Yesterday morning on the way down I came across a big rattler asleep. I cut him first in the back with the sharp edge of the pick, then cut off his head, then his rattles. He had ten rattles so must have been quite an old one. We saw one the day before, but he ran down a hole. They seem rather common but generally give plenty of warning with their rattle, and also try to get away. We also saw a coyote Tuesday afternoon.

Thanks for the bird book. It confirms my identification of practically everything, even the Wilson Phalarope. I saw a flock of Cowbirds the other day, the only new one for some little time. Lark Buntings are abundant here on Hart Table.

We will work this area a little longer then probably go to the vicinity of Sheep Mountain. Our good area near here seems to have become worked out.

I think I will send Aunt Alice some seeds of the most attractive flowers of this country, the blue foxglove, the larkspur, and yucca, and the mariposa lily, also perhaps the white thistle-like poppy.

I guess I will close now as there seems to be no more to say.

Lots of love, Harold

Reference

Scott WB (1913) A history of the land mammals of the Western Hemisphere. The Macmillan Company, New York, 786 p. Illustrated by R. Bruce Horsfall

Abstract

They continued collecting south of Hart Table. Botany professors from Carnegie Institution, Yale and University of Arizona arrived at their camp along with one wife. They found lodging and food in Scenic to be intolerable so came out to the camp. Harold noted that the rudely treated wife basically did everything from driving to making up the camp and arranging the mirror for shaving. Wanless described the Carnegie Institute leader as "too much of a scientist and not enough of a man." Wanless continued into the Indian Creek area describing the sequences, collecting specimens, and interacting with and better understanding the farmers and locals. They began building and filling boxes with wrapped and slightly prepared specimens to ship back east.

Keywords

Indian Creek · Botanist visitors · Shipping containers

Friday July 23

It was cooler today again and quite endurable. We went back to the *Entelodon* peak area and looked over part of it again. We specially wished to see what was in the higher beds here as they were present in good exposure further up the canyons. We found that the upper (green) nodules above the brown ones contained very little and it was not worthwhile to prospect them. Mr. Sinclair found an *Entelodon* skull (rather poor) which we had missed, and I found a good *Oreodon* skull not seen before (front teeth and lower jaw complete). We worked up toward the head of two of the canyons and the second one in the afternoon ascended to the gap at the south end of Hart Mountain. Here we worked along a ridge and I went up on a table south of this which I thought to be a part of Sheep Mountain, but which turned out to be only an isolated flat-topped butte of about an acre's area. The grass was matted thick and apparently no one had ever been on it. It is just about half a mile north of Sheep Mountain and has very steep slopes on all sides. We worked down this canyon again and found a lower jaw of another *Entelodon* which we collected, making quite a load for the day. We started back to camp at about 3:30 feeling that we had at least cleaned up the *Entelodon* Peak area quite thoroughly. This was my last visit to it.

When we got back to camp, we found that Professor Clements of the Carnegie Institution, a botanist, Professor Wieland of Yale, a paleontologist, and Professor Voorhees of the University of Arizona, a zoologist, and Mrs. Clements had called on us. They came back after supper bringing with them a treat of ice cream. This party is going through the badlands of the United States very rapidly with the idea of working out the flora of the badlands and the fauna, in an attempt to see if badland conditions might have influenced the flora and faunas of Tertiary time. They had worked in Sioux County, Nebraska, and intended to go to Wyoming (Bighorn country), then to Colorado, New Mexico, and Arizona. Mrs. Clements drives the car and tends to all the mechanical details of the party. Mr. Clements had found the hotel room in town "Frankly impossible", as he had told the hotel keeper and the restaurant the same way, so had decided to come out and camp on the table near us. Mrs. Clements made up a bed across the seats of the car, and then Mr. Clements inquired if she had it lighted satisfactorily and if a mirror was fixed for his shaving in the morning. We laughed much later about Mr. Clements as he shows himself to be a poor westerner in not being able to make himself agreeable to the people with whom he comes in contact in the field. He is too much of a scientist and not enough a man. The party proposed to drive up to the head of Corral Draw the next day if possible. We are doubtful of the possibility but did not discourage them entirely.

Saturday, July 24
Our friends the Clements' party went into town early as they had to have bacon and eggs for breakfast. Mr. Sinclair and I started out on an exploring expedition to the region of the cedar covered butte to the southwest of camp. We followed the Malloy's road down the canyon below camp and to the point of branching of the forks beyond Leo Malloy's black shack. This part of the valley of Indian Creek has many buttes that are high and of which the lower 60 of 70 feet is in the Pierre Shale, the marine shale of the Upper Cretaceous. An interesting observation here is that the slopes of these buttes are in general grass covered up to the sharp line separating the Pierre Shale from the *Titanotherium* beds (Fig. 10.1). Above this, they are almost entirely barren of vegetation. The bright color of pink and green of the *Titanotherium* beds is here very noticeable. We turned off the road about a mile above the forks and worked through the *Titanotherium* beds to the base of the butte visible from camp with quite a number of cedar trees on its side. This butte rises to about the height of Hart Table. We

found the nodular layer exposed on the side of this butte in quite a good exposure with a good deal of scrap. We worked almost all the way around it and somewhat along the rough area of high badlands connecting this with a similar butte about a mile to the south, which, however, had no cedar trees. I found the dentition of a horse which was collected, but nothing else collectable. We found an interesting structure in a vertical dike of hard sandstone which was about 1 or 2 feet through. This weathered more slowly than the surrounding clays and so many fallen blocks of it were visible. On either side of it was a thin line weathering bright red on the surface. We were not sure whether this side indicated the extent to which infiltration took place or had some other meaning.

We continued on to the south about a mile to the base of the high butte to the south of the cedar covered butte. Beyond this it appeared that the nodular layer changed to a solid sheet of caliche. We started back for camp at 2:30, arriving at about 5:30. We passed the tent and shack of the Malloys, which is the farthest up Indian Creek and stopped to talk to Mrs. Malloy. She is a very good-natured Irish woman and very talkative. The Malloys are what Mr. Sinclair considers professional homesteaders. All the members of the family seem to take up as much land as the law will allow and then as soon as it is proved up on, sell out to cattle men. Tracts of land between camp and their present homestead have been taken up by two sons, Leo and Dan, and sold to Earl Taylor after they have been proved up on. The old people expect to sell this land to Earl Taylor as soon as possible. We were quite tired when we returned to camp as the distance covered had probably been about 20 miles. As rain was threatening, we did not go into town. It rained quite a bit at night.

Sunday, July 25
Today I did my week's washing in the morning and wrote two or three letters. Rather early I went down into the valley of Indian Creek on the road to the Malloys to take pictures of the badlands of the *Titanotherium* beds. The *Titanotherium* beds consist mainly of grayish sandy clay, but in many

Fig. 10.1 The valley of the east branch of Indian Creek opposite Leo Malloy's black shack, a typical badland valley. The bottom, which is in the alkaline Pierre Shale gumbo, has sagebrush, cottonwoods, snow on the mountain, sunflowers, grass, and other plants, while the slopes above the limit of the Pierre Shale are grassless, having a large flowered white primrose as the commonest form of vegetation. A few cedars occasionally grow on the badland slopes. The buttes in the picture are in the *Titanotherium* beds. (Color autochrome photograph taken 25 July 1920)

places contained channels of heavy cross bedded sandstones. I photographed two of these channels, which are shown in Figs. 7.3 and 10.2. I found an exposure of *Titanotherium* beds on some buttes south of Leo Malloy's black shack which contained a great deal of *Titanotherium* scrap material. I looked this over a good deal but found nothing collectable in it. In many places the surface is fairly littered with the huge bones. In the valley of Indian Creek, the unconformity between the *Titanotherium* beds and the Pierre Shale is seen frequently (Fig. 10.3). The beds above the contact are frequently bright shades of reds and browns for some distance up, gradually changing into the gray of the middle and upper *Titanotherium* beds. I returned to camp by the saddle trail following the divide between the branches of Indian Creek going west toward Bowen's place, and those going east to the branch heading at Taylor's spring. This was an interesting trail as it gave a broader view of the Badlands than can be had from following up the valley, and when the sun broke through the clouds, a little before sunset, the long shadows from the low badlands produced a wonderful effect, which is shown in Fig. 10.4. I reached camp by about 7 o'clock in the evening.

Monday, July 26
This morning we went into town quite early as we have not been since the Wednesday before and were almost out of provisions. We carried with us the big *Entelodon* skull, and Mr. Sinclair, in the back yard of Mr. Bump's store, built two boxes in which he shipped all the material which we had collected so far except the big turtle. These required the greater part of the day to make, as we took apart old boxes and with the material in them made new ones. We had lunch at Mrs. Taylor's. Our box making furnished a good deal of interest to the natives so that at many times during the day, quite a gallery was watching the process and commenting on the big skull and other of our finds. Their usual query was: "Now, what kind of an animal do you suppose that was?", and "How long ago did they live?", and they generally had information as to having seen some awfully big bones somewhere a few

Fig. 10.2 Torrential cross bedding in the sandstones of the *Titanotherium* beds on the south side of the east branch of Indian Creek. This cross bedding is quite typical of the channel sandstones of the region. (Color autochrome photograph taken 25 July 1920)

Fig.10.3 A large butte at the main fork of Indian Creek, of which the lower 100 feet of so is in the blackish gray, marine, saline Pierre Shale, the middle section consists of the *Titanotherium* channel sandstone, and the upper part is the *Titanotherium* clays. This is an excellent illustration of the Tertiary – Cretaceous contact. On the creek bottom in the foreground, may be seen numerous prairie dog holes, and also their conical sand hills which reach a height of 8 inches of more. (Color autochrome photograph taken 25 July 1920)

Fig. 10.4 Late afternoon view of the big badlands of Indian Creek valley from a point about half a mile south of Hart Table, and southwest of camp, showing Sheep Mountain to the left, and the high butte dividing Indian Creek and Corral Draw drainage on the extreme right. The Badlands of the foreground are in the *Titanotherium* beds. (Color autochrome photograph taken 25 July 1920)

years before ("titanothere scrap"). The people are very interested in this work and most of them have come in contact with previous collectors, as this is a famous collecting center. Mr. Clements was in town for a short time. He had been unable to get to the head of Corral Draw and had returned to Rapid City to spend the night, thinking no place in Scenic was "possible?" [Almyr] Bump, the son of Mr. Bump the storekeeper has just graduated from high school and is very bright and interested in geological work. He intends to go to college after a year's work. The boys who run the Rochdale store are college students, one being from the University of Idaho and the other from Ohio State University. We returned to camp and did not do any prospecting today.

The Past: The Key to the Present, the Longest Hike July 27–29

11

Abstract

In his hiking Harold observed the modern fauna of coyotes, rabbits, and prairies dogs have close affinities to the ancient Oligocene predator-prey fossils. In extending his prospecting south and west from Hart Table, he was hiking as much as 30 miles a day with limited water, having to rely in part of muddy water pockets and the friendly farmer families. Letters to Harold's mother show the differing modes of maintaining effective planning, communication, photography, and shipping a century ago.

Keywords

Thirty-mile hikes · Predator-prey relationships · Communications

Tuesday, July 27

This morning, as Mr. Sinclair wished to take one more look at the *Entelodon* Peak area and thought he could do it alone, I started out rather early. Early in the morning there was a very heavy fog so that it was impossible to see even Mr. Motter's house at a distance of half a mile. The badlands were not visible at all from camp. This soon rose and it was a fine day for hiking, a little cool with a fine breeze. I started out on Hart Table and fol-lowed the "skyline" trail on which I had travelled Sunday for about a mile and a half, then turned down a branch of Indian Creek. I saw a large bull snake but did not kill it as they are harmless. They are frequently larger than rattlesnakes and have pointed tails. I also saw a coyote standing on the edge of a prairie dog town where several prairie dogs were barking loudly. The coyote retreated a little distance up the hill until I was out of sight and then returned, no doubt, to his prey. On the Saturday preceding, Mr. Sinclair and I had intercepted the pursuit of a coyote of a cottontail rabbit. These incidents merely served as an inside look into the wildlife of the Badlands. It is interesting to contemplate the small change which has taken place in some ways on the earth in a hundred thousand years or more.[1] At the time of the deposition of the clays and sandstones forming the modern Badlands, the vicious *Hyaenodon* and the sly sabre-toothed tiger and the small *Daphoenus*, the ancestor of the modern coyote, no doubt stalked and pursued the swift *Palaeolagus* (ancestral rabbit) and tried to capture the burrowing *Ischyromys* (an early rodent).

---

[1]We now interpret the rocks to be about 32 million years old based on radioisotopic dating (unavailable for sedimentary rocks in 1920).

Now the thousandth or hundred thousandth descendent of the Oligocene carnivore, the coyote, may be seen in the same area pursuing the modern cottontail and jack rabbit and stalking the prairie dog.

After I had crossed the valley of Indian Creek, I worked up the low badlands west of it to the table which separated the lower part of Indian Creek from Little Corral Draw drainage (Fig. 7.1), as Hart Table separated Indian Creek from Spring Creek. At first, I found no exposures of the nodular layer but later found an excellent development of brown nodules where a branch of Corral Draw cuts almost through the table. There were many signs of scrap here but very little collectable material which I discovered. Here most of the bones are weathered a rusty brown instead of being white as were those of the *Entelodon* Peak area. After prospecting this area until about 2 o'clock in the afternoon I struck off south in the drainage of Little Corral Draw. It was interesting to note that the badlands seen from this table seem to be as endless in extent to the west and south as those seen from the edge of Hart Table. I went south from here for about 2 miles for a point west of the cedar covered butte mentioned above and then worked up and down a number of canyons of Little Corral Draw, till I struck the main east branch of this stream. This brought me soon to a point at which the nodules were favorably exposed. The nodules here are both more numerous and larger than in the *Entelodon* Peak area of Indian Creek. There are great fields literally covered with the immense brown nodules of the red layer. There seemed to be comparatively little material here, although turtles were common and other bits of scrap proved it to be the equivalent of our other exposures of the red layer. As it was getting rather late, I was able to spend only an hour in this area but was convinced that the great development of nodules would merit further prospecting. I started back for camp at about 3:30 in the afternoon, stopping for about 10 minutes at the Malloy's tent to get a drink and rest. I had become very thirsty during the day as I had gone at a fast, tiring pace all day and drunk some quite muddy water from a mud hole in Corral Draw. I hiked back from the Malloy's to camp in a little over an hour, reaching camp by 6:30. This was my longest hike of the summer, as I covered probably 27–30 miles, hiking steadily from 8 till 6:30. Mr. Sinclair was much interested in the new nodule field and decided that we would again prospect it the next day. He had found some interesting data in the canyon west of *Entelodon* Peak in the way of algal balls and cakes, which goes to prove that algae form the basis for some of the badland deposits. He says these are perfectly typical algal balls such as are seen in the New York Paleozoic section, and that they indicate the presence of damp meadows.

Wednesday, July 28

This morning Mr. Sinclair and I took the Malloy's road up to the head of Indian Creek and went to the pocket of nodules which I have discovered on Tuesday. We arrived there at about 10 o'clock. We worked the Corral Draw side in the morning and found hosts of nodules but little or no material in them. We decided that this section had been collected in rather recently and, as it has been one of the famous collecting spots for the last 30 years, it is not strange that much of the material should be removed. Hatcher collected a great deal of material from Corral Draw. We crossed back to the Indian Creek side a little after noon and prospected in it for a couple of hours in the afternoon. It was very hot and out water gave out by about 1 o'clock and so we soon got almost parched with thirst. We found a somewhat broken *Entelodon* skull which we left, and a perfect *Oreodon* skull which we took. The rich nodule exposure seemed to extend off indefinitely in the direction of Sheep Mountain, but we did not prospect much of it. We started back at about 3 o'clock and got a drink at the Malloys, which was very welcome and then went on rather slowly back to camp. We were both quite tired, I mainly from the hike of Tuesday and [for] Mr. Sinclair mainly from this day's hike. He considers the area at the head of Indian Creek interesting but too far from our camp site to be easily worked. We arrived at camp about 6:30 in the evening.

Thursday, July 29

Today we went into town in the morning and got our mail, some supplies, and had our hair cut, which was badly needed. We then returned to camp and took a rest the remainder of the day. We did, however move camp about 50 feet north as the place around the tent had become so dusty that the dust and dirt could not be kept out of the tent. We were both quite tired and stiff from the hiking so far this week. I wrote two or three letters in the afternoon.

Scenic, July 29, 1920.

Dear Mother:

I have now gotten three letters from you since I last wrote, although my last letter to you was mailed only three days ago. Last week, we did not get into town from Wednesday till the following Monday so, although I wrote you last Thursday, it did not get started until Monday. Last Thursday was terribly hot but it turned out a bit cooler toward night, and the hot wave was turned. It has been reasonably comfortable since then, although hot enough in the sun. They say that last Thursday it was 115° in the shade in Scenic, and we were not much in the shade, so you can see what we enjoyed for a couple of days.

Saturday, we went off to a cedar-covered butte about 6 or 7 miles from here near the place being homesteaded by the Malloys up near the head of Indian Creek. It was a long way and we were pretty tired by night. We found a fairly good exposure of the nodular layer there but did not find anything extra good. Sunday, we had as a day off, but in the afternoon, I went out and took a few pictures in the nearer badlands. Monday, we went into town and made boxes to pack up what we have collected so far. They filled two very good-sized boxes weighing 150–200 pounds each. They were all carefully wrapped and packed with excelsior, so I am pretty sure they will get there all right. This took most all day, so we had lunch in town and did not do any prospecting.

Tuesday morning, Mr. Sinclair went back to work over some of our old area around where we found the Entelodon skull and I went out to prospect a new territory. I followed a saddle trail down a comb ridge from near camp down into the valley of Indian Creek about 3 miles southwest of here, then worked up a table dividing the drainage of Indian Creek and Corral Draw. Here I found a quite good exposure of the nodular layer which I spent 2 or 3 hours prospecting. I found a few skulls but none rare or perfect enough to take. I then worked down one branch and up another of Corral Draw, till got up toward the head of the main eastern branch. The head of Corral Draw is the region from which a great many of the best things in the Princeton and other collections have come from. Mr. Sinclair had intended to go to and camp near the head of it till he found that it was absolutely inaccessible by car. [A] saddle horse or light wagon would be all that could reach into it. He expects to come into this country again with a light wagon (probably next summer) and will then probably work this area to some extent. It appears that the section we were in had been worked by someone else or else was comparatively poor. As [this was] the richest exposure of nodules we had yet run across, we expected to find a good many excellent things. I returned from this area to camp pretty well tired out and had walked steadily from 7:30 a.m. till 6:30 p.m. with only two short stops, one for lunch and one at the Malloy's place on the way back. I figure I walked 25 to 28 miles. This was the furthest yet this year. Yesterday, Mr. Sinclair and I went back into the same region but did not have very great success. We found a lot of nodules but no skulls. We collected only one good Oreodon skull with lower jaw attached. Mr. Sinclair was badly tired out by the one day there, so he decided to do no prospecting today. We went to town this morning and got supplies and haircuts and I sent a draft for $4.00 to send me more plates. We moved the tent when we came back as our prairie was getting rather dusty. We only moved about 30 feet, but it was quite a job as there were two tents and all their contents to move.

We expected to go to Sheep Mountain tomorrow and prospect a bit in that area. That is the most rugged part of the whole Badlands and is about 8 miles from our camp site.

Now I will try to answer a few of your questions: (1) as to how I sleep. I have slept fairly well but the mosquitoes are frequently bad, and I have stayed awake all of one or 2 nights with them. When it is too cold for them, I generally sleep quite well but am waking earlier (4 of before) when the flies start to bother me. Their tent can be fixed so as to keep out mosquitoes or flies but as mine adjoins the car the only way is to drape mosquito netting over my head which I can hardly stand, as it has to be draped over a pick handle and if I move much I knock it down and have it over my face. However, I get enough sleep to do.

I have done my washing the last two Sundays, which is the day we have off. I sent my wool khaki shirt to the lady in town who is doing some washing. Mrs. Sinclair washes out the lighter things for herself and Mr. Sinclair and sends her heavier things to the lady in town.

I get along nicely in the way of food, and Mrs. Sinclair gets us very good meals. I generally help with the dishes. Mr. Sinclair makes fun of me to some extent for not eating meat and tells her to fry a tuft of prairie grass for me when they have meat, and also advised me to get over the habit as it would never do me any good, but as yet I have not indulged in any. They have very little meat as it is hard to get here. Mr. Sinclair is not at all particular about what he eats. All he cares is that he "stuffs down" enough food as he says. He will never offer any suggestions in the way of meals but lets Mrs. Sinclair prepare everything.

I think one trouble with Mr. Sinclair is that he has no desire for recreation. He never goes to any form of athletic contest at Princeton, as he says he understands nothing about it and cares less. He does not care at all for the theater drama. He does not care for reading stories, nor playing games of any kind. When we take a day off, he usually reads the paper for a little while and then lies down and sleeps the rest of the day except at meal time. His sole recreation is tinkering with the Ford, changing parts, and doing little repair jobs. He is very careful with the car and does not wish anyone else to drive it but himself, so I will probably not get a chance to learn. He has had it since 1916 and it is still in very good shape. I hope you will not take this as very critical, as we all have out faults, but I merely mention it as you might like to know a little about what kind of man I am working with. Even driving cars is not a recreation with him, and he never takes a ride just for a ride, but rather as a means of travel from one point to another.

Mrs. Sinclair is very different. She is very careful about meals, always trying to have good things and well-balanced meals. But one thing which always displeases me is that she always insists on feeding us first and saving only a sixth or so of the food for herself saying she does not care if she has any of not. It always seems as though she is trying to starve herself. Mr. Sinclair is a very rapid eater and is generally almost through a meal by the time I am started and before his wife is started. He never shows any appreciation of food but "stuffs it all in" as he puts it. I fear I have become too critical and hope you will not take these remarks as too much that way, but it is just *a result of a sort of a desire for a real companion, which I feel that I lack here.*

Now, as to your sending some berries, I guess I said in the last letter that you had better not try it, but I think if there are any left when you get this, it might pay to try. They would be a great treat at least to me if they came through. We have had one of two sour green apples from the store. They are one thing Mr. Sinclair cannot eat under any circumstances unless cooked, as he says, "apples do not agree with him." It sure makes my mouth water to hear of all the fruit you tell about. There are a few bushes of an insipid tasting wild red currant which I ran across last Sunday and which I ate eagerly, so you see I seem to have a great desire for wild fruit.

There was a party of botanists in this section two of three day ago who camped one night near us. Professor Clements of the Carnegie Foundation desert Laboratory at Tucson, Ariz., Prof. Weiland of Yale, and a Mr. Voorhees, a geologist of the Carnegie Foundation. These people did not make a very good impression in Scenic as they engaged a room at the hotel, then later looked at the room and told the proprietor that the room was frankly impossible and that they would rather camp out. They did about the same with Taylor's Restaurant. When they camped near us, Mr. Clements came over and got to talking scientific matters with Mr. Sinclair. When it was suggested that he might want to get their camp ready for the night, he said that he was not the mechanically inclined member of the party and that Mrs. Clements ran his car, took care of it, did his type writing and other jobs while he furnished the brains – or words to that effect. People of that sort never do make a good impression, I guess. When Mrs. Clements had the camp all fixed and bed made, he told her she better put up the curtains and fix things so he could shave in the morning. "What a man!" Mr.

Sinclair has the advantage of being a good mixer with the natives which the other man lacks.

You were worried about my getting lost. Please don't worry as the kinds of getting lost in the badlands are not half as bad as they are in the woods right in Lakewood [the family's cottage in Michigan]. When you have been in a section a day or more, you have spotted a half a dozen or more very good landmarks which can be seen for several miles distinctively. Sheep Mountain, our Entelodon Peak, the gap at the end of Hart Table, the Cedar covered butte, and many other landmarks are of this sort.

It has started raining a bit, and I have the tent closed up quite well, so I think I will finish the letter. Thanks for the glasses. I have not yet used them as I accustomed my eyes to the stronger light but will use them some time I think. The colored photograph of the cottage looks very natural, makes me want to be there.

My ear is about the same, does not pain me any, and once in a while clears up a bit. My arm is also a bit better.

I think I can get along all right in the way of clothing with what I have unless you have a pair of slippers or light shoes to put on nights. My shoes are being fixed and I am wearing some heavy gunboats of Mr. Sinclair about 2 sizes too large for me and it would help if I had something to change into to rest after hiking 20 or more miles. Also, these overalls would help to change with my white senior ones. I can do well enough with the shirts I have. They have enough wear in them for this use.

I am sorry the flowers were not in good shape. The big one is the cactus; the mariposa lily is much smaller, and I had none to put in as I neglected to get any the first thing at Interior, and they were done blooming. The larkspur and fox glove are now also done.

I killed or at least seriously wounded 2 rattlers last Friday morning along the edge of the table south of here.

We expect to start back the first of September or thereabouts, and I will travel as far as Chicago with Mr. Sinclair unless our present plans are changed. We are quite likely to go by Rapid City and the Black and Yellow highway[2] which follows about the route by which I came out on the train.

I must close now as I have made this letter about long enough for one attempt. Have referred to your letters and tried to answer most of the questions you have asked.

Lots of love, Harold

[The following letter to Eugene is a follow up pleading for development of the color plates.]

Scenic, July 29, 1920.
Dear Gene,

As we are going into town soon, I want to scratch off a bit of a note to you. I have been too busy and too tired nights for much writing but will have a lot to tell you when I get back.

I wish you would send me 3 dozen more plates (Standard Orthonon 5 × 7) packed well enough to stand a bit of rough usage. I will enclose a money order to cover what I estimate it will cost. If it's more I will settle when I reach Chicago again.

I have not yet heard from you about the autochrome plates but hope to soon. Have only one more exposed now. Will take to other three soon and send them. Hope to hear from you soon as to how the autochrome plates came out as I am rather waiting the others to hear about them. Will you send the plates as soon as possible, as I have only about a dozen left now and want to do a good deal more photography.

It's time to go, so I must stop. Harold Scenic, Pennington Co., S. D.

---

[2]Probably a paved road roughly paralleling Interstate 90.

Abstract

In a solo prospecting up a sheer, narrow, 300-foot-high canyons of Sheep Mountain, Harold describes a quite hair-raising climb as he worked up and down the pinnacles. The following days with Professor Sinclair, they began collecting in one of the most productive fossil localities of the summer, the Bear Creek Pocket. They collect giant tortoises, oreodonts, *Hyaenodon*, *Poebrotherium* (ancestral camel), *Dinictis* (a small sabre-tooth tiger), and *Hyracodon* (a small relative of the rhinoceros) skulls as well as a complete *Oreodon* skeleton all within a few days. He describes the gracious interaction with the local folks and the challenges of driving on the wet, muddy roads.

Keywords

Bear Creek Pocket · Giant turtles · oreodonts · *Hyaenodon* · *Poebrotherium* · *Dinictis* · *Hyracodon*

Friday, July 30

This morning Mrs. Sinclair was not feeling well and, as I did not wish to stay in camp again, I went down Hart Table and into the badlands of Indian Creek at its end. I then hiked up along the base of the northwest corner of Sheep Mountain to where we have last left this section on the very hot Wednesday the 21st. I found a fair exposure of the nodular layer with considerable scrap material and many turtles but nothing collectible. The nodular layer soon ran under the prairie level and was only exposed in stream beds and there not favorably. I came to a large canyon leading up to the pinnacles of Sheep Mountain and ascended it (Fig. 12.1). A number of cattle which were up the canyon came charging down toward me but shifted to the side and galloped away out of sight. The canyon was quite easy to ascend for some distance but became steeper and narrower with vertical walls [coming] down in all directions. I ascended this canyon till I was stopped by a vertical drop of about 20 feet which I could see no way to pass, so I was forced to start down the canyon again. A little down the canyon was a branch which had worn a very steep talus slope with an angle of about 60° or more. I dragged myself up this by sticking in the pick ahead and pulling myself up to its level then advancing it and drawing myself up to it, etc., till I had gotten about 100 feet above the canyon level. Here there was a small vertical section and in it I tried to go up a tiny branch of the canyon, which was so narrow, however, that I could not squeeze into it. I managed to get up a bit higher and then pulled myself up over a practically vertical face of about 6 feet by cutting places for feet in the rock. The rock here is all *Leptauchenia* clays. It breaks very irregularly and has in general very vertical weathering properties (Fig. 12.2). I finally reached the

Fig. 12.1 Gateway to the School of Mines Canyon, Sheep Mountain, showing the *Leptauchenia* and upper *Oreodon* beds. The *Leptauchenia* beds are capped by the White Ash layer and are weathering into vertical columns. (Autochrome photograph taken in 1921)

top only to find myself on an isolated pinnacle about half a mile south of the main mountain and separated by the gigantic canyon of one of the main east branches of Indian Creek, which is here about half a mile wide and about 600 feet deep. As the tendency of all the upper sections of the canyons is to have a gentle slope for some distance before they make a sudden leap of one or two hundred of more feet, it was fairly easy to work along some of the knife edge divides by going up and down the heads of a number of side canyons. In this way, I reached another isolated butte which is the highest point on Sheep Mountain in order to see where I could stand the best chance of getting either to the main part of the mountain or to the river valley of Indian Creek. It was very hot, and I was very thirsty but as I had only a quart of water left, I refrained from drinking and or eating lunch until I should get to a position that I was sure of getting out of. There are three or four immense canyons coming in from Indian Creek to this part of the mountain as well as two or three branches of the White

River to the south (Fig. 12.3). I worked up and down the heads of a number of side canyons of the south branch of the main Indian Creek canyon, hoping that one would have a gentle decent to the canyon bottom, but all of them had vertical drops of 100 or so feet. I was about to let myself down over a seven-foot drop, in a very narrow part of one canyon, but stopped, knowing it would be either impossible or very difficult to get up if I let myself down. I found later that there was a drop of 100 of more feet below this, so it was wise I did not take it.

After heading about two dozen canyons, I reached a part of the mountain with much grass and many red cedar trees and pines,[1] which was larger than any part I had yet been on and followed it for some distance. I soon found that a trail which led about half a mile to the east to a divide between the head of the big canyon of Indian Creek and a large canyon flowing into

---

[1]Now known as Cedar Butte, just to the southwest of Sheep Mountain Table.

Fig. 12.2 Looking down the Grand Canyon of Sheep Mountain from a promontory in the southern section of the mountain, showing the wonderful vertical weathering of the *Leptauchenia* volcanic clays. The shadow in the distance is on Hart Mountain, and beyond that is seen Hart Table, 10 miles distant. (Autochrome photograph taken 4 August 1920)

Fig. 12.3 An immense canyon in the west of Sheep Mountain, showing the upper two hundred of so feet of *Leptauchenia* clays (volcanic ash), near the place where I climbed a similar canyon July 30th. It was on the cedar covered area in the middle distance that I thought for some time that I was stranded. Note the absolutely vertical weathering of the pillars in the left foreground. (Photograph taken 4 August 1920)

White River. There was a trail across the divide and soon I arrived at the main part of the mountain at about 2:15, after fearing for about three hours that I would never leave Sheep Mountain. I had lunch and a drink of water and started enjoying the stupendous canyons on all sides of the mountain. I hiked along the road from this point (Stony Pass) to Hynz's[2] place near the north end of Sheep Mountain Table. Mrs. Hynz is an old German lady but is very pleasant and talkative. She gave me some water and a glass of milk and I rested there for several minutes. She was much interested in my climbing difficulties and said a good many others had gotten in dangerous positions there. She said that when the Schools Mines students came there, they used ropes to let themselves down into the School of Mines Canyon. I went from the Hynz's place north to the road Mr. Hynz has been cutting in a canyon at the north end of the mountain. A few cars have made it to the top this year on this road, but it is quite steep as the height of the mountain is about 300 feet above the creek level at this point.[3] I then worked along the base of Hart Mountain to the gap at the south end of Hart Table, and then back to camp arriving at about 5:30. This had been a day of many experiences but of no collecting. [A letter of this July 30th adventure was written to Harold's mother two day later; see August 1.]

Saturday, July 31
This morning Mr. Sinclair and Mrs. Sinclair and I all went in the car to Scenic and then south on the reservation road for about 4 miles, where we found something of an exposure of brown nodules right in the road. We got out and worked over this exposure for some time, but I found nothing collectable in it. We then ran the car over to the edge of a high butte west of here on the divide

between Spring Creek and Bear Creek drainage. Here we left the car and worked north. There were a great number of big concretions of chalcedony with beautiful botryoidal centers here. We took two of these. We soon became convinced that turtles were thicker here than we had seen them anywhere else, though we had thought them as thick as they would be found anywhere in some sections of the *Entelodon* Peak area. They were here of all sizes from one or two pounds up to 200 or more pounds with rather a predominance of the larger sized ones. At about 10 o'clock Mr. Sinclair found an excellent *Hyaenodon* skull with fine white bone and part of the skeleton. At about the same time I found an *Oreodon* skull and lower jaw in a hard nodule, but strangely the lower jaw was above the skull instead of below it, showing some washing or shifting before petrification. As we had already found two perfect small turtles, we returned to the car with this load. Mrs. Sinclair returned with us to our new pocket and we located two or more perfect *Oreodon* skulls before eating lunch.

Early in the afternoon I explored the section west of this, closer to the middle one of the three high buttes south of Scenic and west of reservation road. Here the white nodular layer at the upper limit of the red layer was exposed right on the surface for the greater part on a flat of several acres. I soon found a rather broken skull of the ancestral camel (Fig. 12.4) which we collected later. I also found most of the skull of a rather small sabre tooth tiger (*Dinictis* sp., (Fig. 12.5) which had both of the sabres present, but somewhat broken off. This was the first sabre toothed tiger with sabres, which we collected.

I went back to meet Mr. Sinclair and found that he had just located a good *Oreodon* skull. We went over together to the area to the west and I found a good skull of the smaller *Oreodon* (*gracilis*) and a minute later a good skull of *Entelodon mortoni* with the lower jaw present. Here we collected all of the stray pieces and as it would require some time in digging out, Mr. Sinclair decided to leave this till Monday. It would be a very good skull as the lower side would probably be perfect. A little later we found another good skull of *Oreodon culbertsoni*, the larger oreodon.

---

[2] Wanless first spelled this family's name Heinz, but later in the diary and in 1921 he consistently spelled their name Hynz. There is no record of a family named Heinz on Sheep Mountain Table in the Pennington County landowner records, but current ranchers recall Happy Hynz and his family living on Sheep Mountain Table.

[3] Sinclair did not drive his car up this canyon in 1920, preferring to park at the base of the canyon. The "road" was probably very bad that summer.

Fig. 12.4 Reconstruction of *Poebrotherium*, the small gazelle-like White River camel. (Illustration by R. Bruce Horsfall in Scott (1913) Fig. 211)

Fig. 12.5 Horsfall's reconstruction of *Dinictis,* the smaller sabre-toothed tiger of White River time. (Illustration by R. Bruce Horsfall in Scott (1913), Fig. 265)

We also located a very large turtle which looked perfect and which would have to be dug out. We left this also for a later time. We started back to town at about 3:30 or 4 o'clock with a big load of material, after looking over a small section east of the reservation road in which we collected almost a complete set of teeth, upper and lower, of *Hyracodon* (Fig. 12.6), the small cursorial rhinoceros. This was the biggest collecting day of the summer. On it we located one *Entelodon* skull, one sabre-toothed tiger, one *Hyaenodon*, one camel skull, one set of teeth of the rhinoceros

Hyracodon, three medium sized turtles, one *Oreodon gracilis* and four perfect skulls of *Oreodon culbertsoni*, a total of thirteen numbered specimens. The people of Scenic were much interested in our discoveries and quite a crowd gathered around the car to see what we had found. I showed them one of two of the *Oreodon* skulls. We then drove back to Hart Table.

Sunday, August 1

This morning I did my weekly washing and wrote some letters. In the afternoon we had quite a call

Fig. 12.6 A reconstruction of *Hyracodon,* a small, gracile relative of the rhinoceros. (Illustration by R. Bruce Horsfall in Scott (1913), Fig. 277)

from Miss Leona Malloy. She was interested in the things we have found and seemed to rather enjoy having us there to call on. Her saddle horse strayed down to the spring below camp, while she was in the tent, and I went down to get him. I mounted him alright, but the horse almost refused to budge in the direction towards camp, so I had to get off and lead him up there. She said he would probably have gone willingly to their camp up in the Badlands but refused to go back. Late in the afternoon Mr. Sinclair and I took a walk down to a near part of the *Titanotherium* badlands to determine if we could define where the line dividing the *Titanotherium* beds from the *Oreodon* beds was. We also found some more fine traces of algal cakes and balls.

Scenic, August 1, 1920

Dear Mother,

   I just mailed my last letter to you yesterday morning as we reached town just in time for me to put it in the mail car of the east-bound passenger train, but the last two days have been quite eventful so I think I will get a little note ready to mail tomorrow. Day before yesterday Mrs. Sinclair

was not feeling well, so Mr. Sinclair stayed in with her and I went on a prospecting tour. I worked along the east side of Indian Creek valley to Sheep Mountain, and, when I reached there, at about ten o'clock, I started up one of the canyons. This led to within about 20 feet of the top, where the rest was vertical. I went down some distance and ascended a side canyon very carefully holding myself in place with a pick while I pulled my body up from point to point. This brought me within a few feet of the top, but after some endeavor I had about decided that it was hopeless till I saw a place I could cut a step with the pick. By this and a good deal more effort, I finally reached the top. I found I was on a narrow ridge separated from the main part of the mountain by about half a mile with a maze of vertical walled canyons in between. I was going to eat lunch as it was noon but, as I was trembling all over from the exertion and danger, I had not the slightest appetite. I decided that I would not eat lunch nor drink more water until I had found a way of getting to the main part of the mountain or down to the valley again. I crossed the heads of a number of canyons by going down into each about 30 feet. I tried each of these canyons and found that each, after it got down a few yards, had a vertical drop of 15 to 100 feet. This was not encouraging, and I looked the problem over. I knew that I might be able to follow the divides around to the main mountain but thought the last divide would be merely a succession of pinnacles. I saw a very steep slope below some cedars which was over 200 feet long and at about a 60° angle. I knew I could probably go down this slow enough by checking my progress with the pick and so worked around toward it. I reached a part of the mountain separated from the main part on the way to the slope and saw cow tracks on it. At this, I decided I must be able to get off some easier way. So I followed this part around to its divide from the main mountain and there found a well-marked trail heading the canyons of that divide and followed it back to the main part of the mountain which is a big table about 3 miles long with a ranch on it. I then sat down to lunch at quarter of 2 with great relief. Sheep Mountain is the ruggedest and highest part of the Big Badlands and has a great mass of pinnacles and vertical walled canyons. I am going to take a number of pictures there on future visits but [I] will promise to be more careful. You see I did get out of a pretty bad hole, so I think you can trust me to do so in the future.

Yesterday, Mr. Sinclair drove to a part of the reservation road south of Scenic on the way to the part of Sheep Mountain I first went to. This is Bear Creek basin. We found a very excellent exposure of the nodular layer there and made the best collection of the trip so far, bringing in 8 skulls, a complete set of teeth and three turtles, besides leaving an Entelodon head, which will have to be dug out carefully, and a mammoth

turtle. The turtles seem so thick in this area that I think we saw between 200 and 300, but only an occasional one was practically perfect. Most of them are enormous things that one person cannot lift them. The big one we are going to dig out will probably weigh 150 pounds or more. The Entelodon I found yesterday is a smaller variety than the big one and is not as perfect as one side is lying exposed and much has weathered off. However, all the pieces are there, and Mr. Sinclair thinks the covered side of the head will be perfect and that it will make a good museum piece. We got a complete set of teeth upper and lower of Hyracodon (a small rhinoceros), 5 Oreodon skulls, a good Hyaenodon (carnivorous animal), and I found a front of a sabre toothed tiger skull with most of both sabres in place. We are going back to this area tomorrow.

I must close for now. Will write soon again.

With love, Harold.

Monday, August 2

This morning we went with the car through Scenic to the reservation road and back to our rich pocket in the Bear Creek drainage (Fig. 12.7) which we had discovered on the Saturday preceding. Mr. Sinclair worked a good deal of the morning in excavating, shellacking, and putting a paste cloth on the *Entelodon* skull which I had found the preceding Saturday. I took the photograph of it in place. I prospected around in the vicinity and found a great deal of scrap material. I dug out a big turtle at about 10 o'clock which was about the size of our large turtle from the *Entelodon* Peak area and carried him over to the *Entelodon* skull. We took this as it was very white and fresh and not rusty as most of the large turtles generally are. In the afternoon at about 2 o'clock, while prospecting, I suddenly came on an *Oreodon* skeleton in death pose which was in excellent shape. It lay in three large nodules right at the surface, and no excavation would be necessary. I

Fig. 12.7 Development of plain of brown nodules by erosion in the valley of Bear Creek near the point where the *Oreodon* skeleton was found. In areas such as this, we had our best prospecting, as many of the nodules contained turtles or skulls and bones of mammals. (Autochrome photograph, taken 2 August 1920)

hurried to get Mr. Sinclair and my camera and tripod so as to get a picture of it as it lay. Mr. Sinclair was pleased with it as the Princeton collection has not a mounted skeleton of *Oreodon* and he thought he could make one from this. I secured two pictures of it and then we dug it up (Fig. 1.6).[4] The pelvis and part of the scapula were missing but otherwise seemed complete. The vertebrae were arranged in sequence, only one having weathered out. It did not take long to get the skeleton up and wrapped in cotton as it separated from the clay below nicely. It seemed remarkable to find anything in as good shape out on the surface of the flat with cow tracks within a foot or two of it. We also found two or more *Oreodon* skulls in this area. We started back to town and camp rather early as it was very hot.

Tuesday, August 3

This morning, we again drove south to the area of Bear Creek, but Mrs. Sinclair stayed in camp. We

dug out the big turtle of Saturday's finding and saw that a few plates were missing from its edge, and also decided that is was too big as the two of us could scarcely lift it from the ground. I took a picture of it (Fig. 12.8) to show the size which turtles reach in the White River Formation. We then prospected a little extension of the Bear Creek area to the south. Here in rapid succession, we found three of four good *Oreodon* skulls, one of which was of the smaller variety. Mr. Sinclair also collected a perfect small turtle and a set of teeth of *Mesohippus.*

Before noon we crossed over a low divide toward the drainage of Spring Creek and found a fairly good exposure of the nodules of the red layer was present there. However, it was not in the same class with the exposure on the Bear Creek side where several acres were exposed in a flat on the level which has proved the most productive, so that one can just walk around and pick up skulls. We found just after lunch another skull of the larger *Oreodon* with a good deal of skeleton material. We also came on quite a good prospect of *Mesohippus,* the three-toed horse, as there was

---

[4]The specimen was mounted and is at the Yale Peabody Museum, specimen VPPU.012565.

Fig. 12.8 A very large turtle which we exposed in the rich Bear Creek pocket [that we were] expecting to [collect], but found it was a little too large for easy transportation. The pick handle gives an idea of its size. The two of us could barely turn it over when we had it excavated. We saw many hundred turtles nearly this size, but mostly somewhat broken. (Photograph taken 3 August 1920)

much skeleton material present. The skull, however, was in such bad shape that we decided to leave it. Further to the south, I found a lower jaw of a sabre-toothed tiger which we collected and one of two teeth of *Cynodictis*,[5] the small dog. About three a rainstorm threatened and, as the reservation road is a bad place to get stuck, we started out in hurry. We managed to reach the road before it started to rain very heavily, but for the first mile or two along this the rain was coming down in sheets. We managed to get along the level stretches of prairie nicely, but on some of the badland creek crossings we skidded so badly that we could hardly keep the car going ahead. We got out of the rain about a half a mile before we reached Scenic, and at Scenic it had not rained at all. We got back to Hart Table all right and had another little shower after we arrived there. This shows something of the local character of the Badlands storms.

Reference

Scott WB (1913) A history of the land mammals of the Western hemisphere. The Macmillan Company, New York. 786 p Illustrated by R. Bruce Horsfall

---

[5]Now known as *Hesperocyon* one of the oldest true dogs (a member of the Canidae). *Cynodictus* is a European Amphicyonid, sometimes called a bear-dog, related to *Daphoneus* of the White River Group.

Abstract

Harold and the Sinclairs drive to the base and then hike up Sheep Mountain to document the sequences and photograph the spectacular sheer and immense canyons using autochrome color plates. Subsequently they focused on and sampled the lower portion of the sequence on the edge of Hart Table near camp. This included the colorful reddish exposure zone of *Titanotherium* (rhinoceros-like) beds overlying the Cretaceous marine Pierre Shale sediments. Heat and intense hiking were beginning to take a toll on all of them, probably enhanced by insufficient water while hiking.

Keywords

Sheep Mountain · *Titanotherium* · Pierre Shale · Cretaceous · Oligocene

Wednesday, August 4

This morning we went in the car to Scenic and then down Hynz's road south from there about 6 of 7 miles to the base of Sheep Mountain Canyon road. We left the car at the base of the mountain and climbed the road up to it (Fig. 13.1). Mrs. Sinclair accompanied us today. This canyon is quite wonderful as almost vertical walls rise 2–300 feet above the road in it. We hiked south along the road and along the edge of Sheep Mountain Table to Stony Pass where the big canyons of Indian Creek and a branch of White River have cut the mountain in two leaving a narrow trail across the divide. The views from the top of Sheep Mountain are wonderful as they show badlands in practically every direction almost as far as the eye can see. The canyons are the most immense I have ever seen, being much wider and more than twice as high as the famous Niagara gorge. The height here above Indian Creek valley is supposed to be about 730 feet. I found a position from the top of the junction of two main branches of the canyon of Indian Creek and there took my first Sheep Mountain autochrome (Fig. 12.2). It was a magnificent subject with a great deal of vertical weathering showing. We had lunch under a cedar tree, which was quite a luxury after so many lunches in the broiling sun. Near here was a very gnarled old cedar tree with its trunk and branches twisted (Fig. 13.2). It may have been hundreds of years old. We went across Stony Pass and right at the middle of it on the edge of a big earthquake crack filling I found a small skull. The *Leptauchenia* clays are very hard matrix and do not separate cleanly from the bone, but we finally got it extracted. It turned out to be a skull of a peccary, of which there seem to be none described from the higher beds. While not a very good skull, it may have considerable scientific interest. We went to the west end of the section south of Stony Pass and looked at some of

Fig. 13.1 View of the canyon at the north end of Sheep Mountain up which Mr. Hynz's road has been cut. The upper 2 or 3 feet are the basal *Leptauchenia* beds, and the rest are in the upper and middle *Oreodon* beds. The valley in the distance is the valley of Spring Creek, or Scenic basin. The road was made by about a dozen cars [driving up and down the canyon] during the course of the summer. (Autochrome photograph taken 18 August 1920)

the immense canyons in that direction. We also placed some of the creeks further [west] in the Badlands, Big Corral Draw, Cottonwood Creek, Quinn Draw, etc., which are all famous collecting places. A storm was working up from the hills, and I wanted to take one more autochrome plate, so I hurried back to Stony Pass although the sun had already gone under a cloud. I took a color plate (Fig. 13.3) and an ordinary plate here. The view showed the distant badlands still in sunlight but the foreground and pinnacles and spires in shadow. We hurried back across Stony Pass and long the road to the Hynz place and had nearly reached this when the rain came. It poured for about 5 minutes, and we got soaked to the skin, but it was soon over. We stopped and talked with Mrs. Hynz and her son and daughter for some time. They are very congenial people and gave us quite a bit of information of interest about Sheep Mountain and the work of former collectors. Mr. Hynz has accompanied Professor O'Hara several times so knows something of collecting methods. We then returned to the base of the mountain and drove back to Scenic and Hart Table.

Thursday, August 5

This morning we decided to work some of the *Titanotherium* beds in the valley of Indian Creek. I took my camera and secured my last autochrome plate of a canyon with the bright colored lower *Titanotherium* beds west of the skyline trail southwest of camp (Fig. 13.4). We found several places in which algal cakes and balls showed to fine advantage and collected some of them. We also found the gravel bed at the base of the *Titanotherium* beds and just above the Pierre Shale where we collected samples. We worked over some of the area of the high *Titanotherium* buttes south of the black shack which I had visited on Sunday, July 24. We found a great deal of *Titanotherium* material and some of a large *Entelodon* and of a large rhinoceros. We also found part of a lower jaw with six teeth in a hard sandstone matrix. We exposed all there was of it, which was about 2 feet long, decided that it was not enough to be of any importance and was also very hard to extract, so we decided to leave it. We returned to camp quite early as Mr. Sinclair was feeling very tired.

Fig. 13.2 Very old, gnarled cedar tree on Sheep Mountain (south of Stony Pass) and probably the oldest tree on Sheep Mountain. Most of the branches were twisted around several times. There is quite a thicket of cedar and pine trees on various parts of Sheep Mountain, and also some wild red-currant shrubs. (Photograph taken 18 August 1920. (Authors' note: These cedars are still there and look very much like they did in 1920))

Fig. 13.3 Looking east from the south side of Stony Pass, Sheep Mountain, showing the spires and pinnacles of the foreground in shadow, and the distant Badlands of White River valley in sunlight. A storm was working up rapidly from the Black Hills, which overtook us before we had reached Hynz's place, 2 miles distant. (Autochrome photograph taken 4 August 1920)

Fig. 13.4 A small canyon of a branch of Indian Creek about a mile southwest of our camp on Hart Table, which shows the transition from the blue gray Pierre Shales at the base through some of the clays colored brilliant shades of orange and red by a secondary infiltration of iron oxide, to the gray clays and sandstones of the lower *Titanotherium* beds. The edge of Hart Table shows in the distance. (Photograph taken 5 August 1920)

Friday, August 6

This morning Mr. Sinclair did not feel like going out, so he stayed in camp. I decided to prospect some more of the section at the head or Indian Creek and so hiked up the valley of Indian Creek past the Malloy's place to where we had left this section before. I found great exposures of the nodular layer everywhere which should be favorable. As I had worn a blister in my heel, walking rather hurt, as I had to limp on it for 15 miles of more. I could find very little except some scrap material of a long time, but a little before noon, I discovered quite a good skull of a large *Hyaenodon*, which I collected along with a good deal of skeleton material after lunch. It was quite heavy as the skull was about twice the size of an *Oreodon* and the animal approached in size the modern black bear. I worked along the heads of various branches of Indian Creek toward Sheep Mountain and found everywhere an excellent exposure of nodules. I found a few low divides separating Indian Creek from Cottonwood Creek which drains into White River and worked for an hour or two in the valley of Cottonwood Creek. I found everywhere there a good deal of scrap material but very little collectable. It seems probable that this was a very rich pocket but that it has been pretty well cleaned out by other collectors. There were some channels of heavy green *Metamynodon* sandstone here, but I did not find any material in them. I finally crossed back to Indian Creek drainage about a mile or so west of Sheep Mountain on a very low divide which had been used by the Indians as a road to Rapid City before the railroad was put through Scenic. I followed this road track down to Malloy's place seeing many range cattle and horses on the way and then followed the Malloy's road back to camp. I was quite tired tonight. The skull was of *Hyaenodon horridus* making five *Hyaenodon* skulls collected, several of which appear to be different from each other.

Scenic, August 6, 1920
Dear Mother,

I have not been able to write since Sunday as I have been out every day and absolutely too tired at night. I am more tired than ever tonight, as I have hiked almost 25 miles, half of it limping with a blister on my heel, but we are going to town tomorrow so I must get off a note.

I received the box on Tuesday, one letter on Monday, and the other Wednesday, so I have a good deal of mail. The box was a great treat. Mrs. Sinclair and I enjoyed the blueberries very much, but Mr. Sinclair did not take any. We have also had the jam on our lunch and toast for 3 days in a row, both very good. Thanks also for the clothes. I will appreciate the trousers as I have had to wear either overalls or military trousers all the time. The mosquito netting will come in handy perhaps, but I had some before. I do not need it every day, and it is uncomfortable.

I have one thing I must get in the letter, and that is a little about going back. We will probably leave about the first of September, and I am to go in the car as far as Chicago. Mrs. Sinclair will go by train that far and has wondered if it would be possible to stay with you. I told her I was quite sure it would be, as I thought you could find a place somewhere. I wish you would let me know right away so I can tell her. This would be about the second or third of September till the seventh or eighth. I hope you can work this some way as she does not want to stay in a hotel, and she does not want another to go all the way back to Princeton for fear she will worry too much about him.

Monday afternoon this week, I had the fortune to find an Oreodon skeleton exposed on the surface in death pose (Fig. 1.6). This is the first skeleton we have found, and the Princeton collection has none of this kind. It is a good find. We found half a dozen or more skulls that same day and also two good turtles. The Oreodon is an animal about the size of a sheep and with some character rather like a pig. It has no modern prototype and there is little known about its habits, but it was the commonest animal in the period we have done most of our collecting in.

Wednesday, we went to Sheep Mountain again and I took two more color plates and yesterday I took my last one. I will send them off tomorrow.

I must close now as Mr. and Mrs. Sinclair are waiting to go to bed. I will try to write more Sunday, day after tomorrow.

Love, Harold

Wild Stallions and Broken Springs, August 7–9

Abstract

Sinclair treated a ranch hand's injury from a bad bite from a wild stallion. Wanless became involved with harvesting of oats and wheat with the local farmers and describes the damage to wheat from red rust and alfalfa from white moths. Prospecting in the heat continues. The wonderful Model-T was a new and invaluable field tool but delicate and needed constant repairs and maintenance.

Keywords

Farming risks · Crop diseases · Model-T Ford

Saturday, August 7

Grain cutting is going on all around us on Hart Table. Mr. Hill has been cutting his wheat just east of the fence and today Mr. Taylor started cutting his as well as his oats. We went into town this morning and packed up a very large box, our second shipment to Princeton. As we did not have enough to fill another box, we did not make one. We returned to Hart Table and rested most of the afternoon. About 4:30 Guy Taylor, a younger brother of Earl, came to camp with his hip badly bitten by a wild stallion which he was attempting to catch in the badlands. Mr. Sinclair cleaned out the wound thoroughly with iodine and found that it went in 2 inches or more. He advised him to go to Rapid City to have it fixed, but as it is both harvest time and cattle shipping time, they decided he could not be spared if Mr. Sinclair could tend to it a bit. Mr. Sinclair and I helped Earl Taylor shock wheat for an hour or two just before sunset. I had had no experience but seemed to get along all right. The wheat is in some places badly rusted with red rust and the grains are much smaller than they should be. This will make the yield only about ten or fifteen bushels per acre.

Sunday, August 8

This morning I did my usual weekly washing, and afterward wrote a few letters. We had a visit from Wesley Taylor, the youngest brother of Earl Taylor. Earl and Guy had gone to town to try to get the catholic father to do something for the wound when Wesley came back from Corral Draw. Wesley is only 17 years old but is very responsible for his age. Last winter he was left in charge of 250 head of cattle, about 25 miles up the Cheyenne River from the railroad, alone. He is probably as familiar with the upper badlands of the Cheyenne River country as anyone. He said that a School of Mines expedition recently camped for over a month in Little Corral Draw, and he thought that its area was pretty well cleaned out. He also told of their finding a complete *Titanotherium* skeleton about 25 miles south of here in Quinn Draw, which had been sent to Chicago, probably the Field Museum. We had quite an interesting visit as he is very well acquainted with the Badlands. Mrs. Sinclair was again feeling quite sick tonight, perhaps as a result of the hot, dry south wind which blew all day very much like that of July 21–22. It turned a little cooler by night, but the wind did not shift to the west till quite late.

Scenic, August 8, 1920

Dear Mother,

I got a short note off to you yesterday morning but as I have more time today and am not quite as tired, I will write a little more.

It has been a very hot day and Mrs. Sinclair is feeling bad again, the third time. I think she is getting tired of roughing it and anxious to get back to civilization. She is getting so that little things disturb her a good deal. Of course, staying in camp is very tiresome and the flies troublesome and the weather hot, but when she went with us to Sheep Mountain last Wednesday and we walked about six miles, she was quite exhausted, so it seemed rather hard for her anyway. Mr. Sinclair administers dope of some kind on all occasions and calomel and aspirin tablets are taken daily or oftener with phenol soda for cuts and other kinds of ills. They use a bottle of iodine every two weeks or so.

After Supper, I am watching the brilliant colors of orange changing to pink. The sun just set west of the Black Hills and the little clouds are brilliant in the west. I did not get a chance to use a color plate on a sunset as I used the last one on the Titanotherium beds day before yesterday and sent them off yesterday.

We packed up another big box yesterday weighing 250 or more pounds containing two big turtles and a lot of skulls.

Yesterday afternoon, Mr. Taylor's brother came up from the Badlands badly bitten by a stallion which has caused a good deal of trouble. Mr. Sinclair stuck a knife over an inch up into the wound and packed it with bandage soaked in iodine.

He is quite lame tonight but will probably have no trouble unless it gets infected.

Mr. Sinclair and I helped Mr. Taylor shock some wheat yesterday afternoon near camp. This was a new job for me, but I liked it quite well. There is a very good stand of wheat on Hart Table but a little red rust in it. The alfalfa seems to have gotten infected with small white moths which are very numerous in it.

I have been reading the Magazine and Book Review sections of the New York Times and found several things interesting.

I will have to stop now as it is almost dark and the lamp chimney was broken this morning, so I guess there is nothing to do but go to bed. We will probably head for a new pocket tomorrow if Mrs. Sinclair is feeling better. If not, I may do some more prospecting myself.

Lots of Love, Harold.

Monday, August 9

This morning, just as we were about ready to start out, we found that the main leaf of the back spring of the car had broken during the night, and consequently Mr. Sinclair said that we could not use the car for prospecting today. He drove into town with the car empty and ordered a new spring from Rapid City and then drove back to camp. I went east across Hart Table to the valley of Spring Creek and prospected a little of the edge of Hart Table there, finding some of the nodular layer but very little material and nothing collectable. I worked along the side of it to the Indian Creek – Spring Creek Gap at its end. I then crossed the valley of Spring Creek finding a few exposures of the nodular layer and finally reached the Spring Creek side of the high butte south of Scenic by the reservation road. I worked along the Spring Creek side of the middle one of these, and in quite rapid succession before noon had found three perfect *Oreodon* skulls with lower jaws and front teeth, all of which I carried to the location of the big turtle in the Bear Creek pocket and left them. I hiked south from the Bear Creek pocket toward Sheep Mountain about 2 or 3 miles finding occasional exposures of the nodular layer but none very good. It was a very hot day. I worked back around the third of the high buttes south of Scenic but found that its base was too high for exposure of the red layer. I then returned along the Spring Creek side of the middle butte to the one nearest Scenic. The nodular layer was here rather high up as all the beds have a gradual dip to the south and east [of] about one degree. I found nothing here and as it was getting rather late, I returned to Hart Table and crossed it just south of Mr. Motter's cornfield. Mr. Sinclair had gone up to the algal cake exposure in the *Entelodon* Peak area and made some more observations and collected a little more material to illustrate this.

Giant Hail Stones, Fossil Dogs, and Sabre-Toothed Tigers, August 10–15

15

Abstract

They retrieved the skulls collected from the rich Bear Creek Pocket, returning to camp for an intense hailstorm with inch and a half hailstones. This severely damaged melons and other crops. Skull collection included *Daphoenus* (an ancient bear-dog) *and Hoplophoneus* (a large sabre-toothed tiger) in new areas north and east of Scenic.

Keywords

Hailstorm · *Daphoenus* · *Hoplophoneus* · Bear Creek Pocket

Tuesday, August 10

This morning Mr. Sinclair again drove into town with the car and, as no spring had come from Rapid City, he used an old one which the garage man Mr. Jost had furnished. He returned to camp about 11:00. I wrote some letters in the morning. In the afternoon we drove through Scenic and up the Reservation road to Bear Creek pocket and gathered my three *Oreodon* skulls of the day before, the camel skull found over a week before, and a camel's foot, which Mr. Sinclair had discovered. We then prospected some of the area for extensions in several directions. We were unable to find any very favorable extensions of the area and so decided that our very rich pocket had about given out. We had made a bigger collection here, Mr. Sinclair said, than in almost any other pocket in his experience.

As a storm was threatening in the west, we started for camp rather hurriedly. It had not yet reached us when we got to Scenic, but, as it was very dark in the west, we hurried on to camp. When we reached Hart Table, the sky to the west was almost black and whirlwinds of dust were to be seen in various parts of the Badlands. We had one tent down, the car shelter tent, and did not put it up but had no more than gotten the car curtains up when it started raining and hailing. It poured for a few moments and then began hailing very hard so that, if one held his hand to the edge of the tent, his finger would surely be bruised by hailstones. These were about the largest hailstones I have ever seen, as some had the diameter of an inch and a half and their average diameter was more than 1 inch. It continued hailing and pouring for about 15 minutes and then cleared up. There was almost a lake all over Hart Table, but it soon ran off or soaked in. We gathered some of the big hail stones and made iced lemonade from them. There was some more rain later in the night, but no more hail.

Wednesday, August 11

This morning it was very cool and rainy, and there was also a heavy fog. It was the coldest morning of the summer. We heard from our neighbor, Mr. Motter, that the hailstorm had done a good deal of damage to the crops, especially melons, which had been almost ruined. We stayed in camp until about 10 or 11 o'clock but started down to the end of Hart Table then, Mrs. Sinclair accompanying us. It was very cool all through the morning. We encountered two big rattlesnakes right together at one point and, as both were coiled for striking, I did not attempt to kill them. We went down into the basin of nodules on the Spring Creek side. I took two pictures of the nodule field on this side and we prospected most of it over quite carefully. I found a set of horse teeth which were worth taking and a little later a fair skull of *Daphoenus*, the ancestral dog, of which we had not yet collected anything all summer. After about 3 hours, we started back for camp. I took a picture of the *Entelodon* Peak area from the south of Hart Table (Fig. 15.1). We did not encounter the rattlesnakes on the way back to camp, although we all prepared to meet them with large stones.

There have been, for several day, flowers of a beautiful variety, large and cream colored which come open about 7 o'clock in the evening. They have a strong musk-like odor and attract sphinx moths in great number. As some of the plants have about a hundred or more flowers, it is very showy and attractive. It grows mainly on the edge of Hart Table as the foxgloves did (this was later identified as *Mentzelia decapetala* of the Loasacae family).

Scenic, August 11, 1920

Dear Mother:

I got the box and letter yesterday. The blueberries were a little soft but in pretty good shape and tasted fine. The shoes are very restful for putting on after a hard day's hike. I have them on now. They would not do for hiking any as the wear on them would be too hard but will be all right for camp.

Yesterday afternoon, we got back to camp with the car when a violent wind came up and in two minutes great balls of hail were coming down. It hailed for about

Fig. 15.1 General view of the *Entelodon* Peak region from near the south end of Hart Table, showing *Entelodon* Peak in the center distance, behind which is the north end of Sheep Mountain Table. To the left is Hart Mountain, and in the foreground are spurs of Hart Table. The east branch of the east branch of Indian Creek, which was used as a road for bringing up the large *Archaeotherium* skull, can be clearly seen. Note the absence of trees over this area which is probably 6 or 7 square miles. (Photograph taken 11 August 1920)

15 minutes, the average size of the hail being about an inch in diameter and some considerably larger. It hurt the melons of our neighbor, Mr. Motter, and some oats he did not yet have cut, but it did not seem to hurt his corn. We stayed in and held the tent from falling during the storm.

Today has been cold and windy and I have worn the woolen shirt and both sweaters. It was raining this morning, so we did not go out until about noon. We went down to the Spring Creek Pocket at the end of Hart Table and reworked an area worked about 2 weeks ago. I found part of a horse skull which we collected and a dog (Daphoenus) skull, the first dog we have yet found. I also took three pictures leaving me only one more plate till the three dozen I sent for from Eugene arrive. I had a note from Eugene saying that only 2 of the color plates were very good. He also said he could send the pictures and that he was very busy.

Our collection now numbers 87 of which 50 or 51 are skulls and 8 turtles, the rest lower jaws and fragmentary material. Of the total, I have found 47, Mr. Sinclair 38, Mrs. Sinclair one, and Mr. Brown of Interior, furnished one by the records, so I guess I have found my share. Monday, the back spring of the car was broken down, so I hiked out alone. I walked about 20 or 25 miles and collected three good Oreodon skulls with front teeth and lower jaws.

We may start back in a little over a week. I will let you know when I know definitely. I would be glad to know from you when you plan to leave Lakewood and whether or not it will be all right for Mrs. Sinclair to stay with you for 4 or 5 days while we are travelling if I go by car.

Must close as it is bed time.

Lots of love, Harold.

Thursday, August 12

This morning we started east through Scenic to hunt for a new pocket. We took the road to Interior for about 4 miles. This is very rough, even more so than the reservation road, though at present it has a great deal of traffic, being the main highway through this section of country. We went through what is called Chamberlin Pass where the railroad crosses the divide between White River and Cheyenne River drainage. We then prospected south of the railroad tracks for some distance. We found an exposure of the nodular layer which was not especially favorable, and which had almost nothing in it. We continued to follow this until it went underground and then turned back over the prairie to the railroad, crossed this and worked up to a few isolated low buttes which contained exposures of the nodular layer. We found nothing in these, so went on into the big area of Badlands about half the size of Indian Creek which appeared to the north. We worked this for some little time, thinking we had discovered exposures of the nodular layer and then had them disappear. We had considerable argument as to the stratigraphy of the beds till, on returning to the car, we found the nodular layer at a level higher than that at which we had been prospecting, with a good deal of scrap material. It appears that part of the reddish clays of the lower part of the nodular layer had been replaced by greenish nodules which resembled very closely those above the nodules, thus causing the confusion. We found a small horse skull in fair shape, which Mr. Sinclair shellacked and left to dry. We then returned to the car. Mrs. Sinclair had found a small skull which she was not sure of near the car, and on examination it turned out to be the brain cast of, and skull of, the ancestral dog, *Daphoenus*, so we collected it. It will be a good addition to the collection, as it is the second *Daphoenus* of the summer. We drove on toward Imlay, about 3 or 4 miles east of here, and I looked around several buttes there where red layer was exposed. The red layer here contained very few nodules and much clay with very little material present. We started back for camp about 4:30 and learned from the map in Mr. Bump's store that the area east of Chamberlin Pass was in the drainage area of Jones and Cain Creeks, tributaries of White River.

Friday, August 13

Today we went east on the Road to Interior but turned off on the prairie west of Chamberlin Pass and left the car out of sight in the edge of the high badlands there behind a small butte. In the valley of Bear Creek near there are lots of stones of flint or chert arranged so as to spell out various names. These are seen for a mile of more along the road. This is probably done by sheep herders having nothing else to do. We worked around from where we left the car down the valley of Jones Creek through Chamberlin Pass and around to the vicinity where we had left the small horse skull. Mr. Sinclair went to finish work on that with Mrs. Sinclair, and I worked up the exposures of the nodular layer in two or three canyons there. The red layer was here exposed in full section with a white band of limey of cherty material at the base and the greenish nodules at the top. We measured the layer's thickness three of four times at different points. Its thickness was about 35–40 feet on average. While I was prospecting up a large canyon, I came on the skull of a fairly large animal in a pretty good state of preservation

but in three of four pieces. I carried as much as I could carry back to where Mr. Sinclair was working, and he pronounced it a very large sabre-toothed tiger of the genus *Hoplophoneus* (Fig. 15.2). It had both sabres present, each of which was probably 3 inches long. This is the best sabre-toothed tiger skull we have collected this summer. We collected the horse and then worked east and north on the Cain Creek side investigating the east side of the high narrow table there. We found that the exposure seemed to grow higher and higher and soon was almost as difficult to work as was the edge of Hart Mountain. We worked along it until after 3 o'clock. I found most of the skull and lower [jaw] of the rhinoceros *Hyracodon*, but it was not of good enough quality to collect. It was very hot, and we soon started back for camp, collecting on the way, some of the skeleton of the big sabre-toothed tiger skull. We then returned to town and camp.

Saturday, August 14

This morning Mrs. Sinclair stayed in camp, and we went back to the same place where we had left the car the preceding day. We then worked north from the point along the base of the buttes south of 71 Table in the valley of Bear Creek. This was the area which I visited on my third day in Scenic. We investigated a large channel in the upper

Fig. 15.2 Horsfall's reconstruction of *Hoplophoneus*, the larger sabre-toothed tiger of the White River, with oreodonts in the background. (From Scott (1913) Fig. 264)

Oreodon beds here and a great exposure of the nodular layer but found very little. It was an extremely hot day. I found the dentition of a horse which we took out, though it was not very good. A little later, we entered a large basin surrounded by large buttes in which a fairly favorable exposure of the nodular layer was found. Mr. Sinclair believed this to be the area in which the first badland fossil hunting was done in 1855–1860 by Thaddeus Culbertson. The locality fitted his description well. We crossed a divide here into the valley of Cain Creek, a tributary of White River which is much deeper cut that the Bear Creek side. We prospected in it for some time but found only scrap material and found it a very difficult section to work, so we crossed back to the Bear Creek side where we collected the foot of a *Caenopus* and a foot of an *Hyracodon*. We started back for camp rather early as it had been exceedingly hot all day.

Earl Taylor made a shipment of 100 cattle today (4 cars) to Sioux City. His father accompanied the shipment. He has been cutting hay on a

Scenic, August 14, 1920

Dear Mother,

I have some good news for you. As Mrs. Sinclair has been getting very tired of camping lately and as Mr. Sinclair thinks he can quit a little early, we have decided to make next week our last week of prospecting. So, Mrs. Sinclair will go east with Mr. Sinclair and I will take an eastbound train probably on the [Chicago] North Western leaving a week from Monday, August 23.

This will get me in Chicago in the morning of Wednesday. I can get the boat Wednesday night for Lakewood and will arrive there the morning of Thursday August 26. . . . . .

I will try to get this letter in the Monday morning mail, but you will get it by

Thursday or Friday. Do not try to send an answer, as I could not get it.

We have been prospecting east of Scenic for several days. It has been poor prospecting, but we have found a few things including the first two dog skulls of the summer and three more sets of horse teeth and yesterday a pretty good find of mine. A very large sabre-tooth tiger skull with both sabres each over 2 inches long. It must have been a vicious animal.

I must close as it is bed time. Hope I will get the box before I go as the last was a great treat.

Lots of love, Harold.

contract with Mr. Arnold way up in the reservation, about 25 miles on Red Shirt Table, all this week. He contracted to cut a thousand tons of hay. We talked over leaving South Dakota and decided we would wind up out prospecting the next week as Mrs. Sinclair has been getting quite anxious to start homeward.

Sunday, August 15

This morning I did my week's washing, the last on Hart Table. I decided I would spend a little time at the cottage in Michigan and wrote mother to that effect. I also wrote seven other letters during the day, and in the afternoon went out for a little while with my camera into the badlands. It was a very hot day. I took several pictures to illustrate the relation of Hart Table to the badlands and also to show the occurrence of the bright colored *Titanotherium* beds. It turned a little cooler by night.

Reference

Scott WB (1913) A history of the land mammals of the Western Hemisphere. The Macmillan Company, New York, 786 p. Illustrated by R. Bruce Horsfall

Abstract

During their final prospecting, sediment sampling, and photography, they encountered chalcedony (silica) veins cutting through the clay strata and continued the sediment stratigraphy descriptions. On August 19, after 5 weeks and 4 days, they packed up their camp and moved back to Scenic where they crated the remaining specimens for shipping east and said their goodbyes to the town folk. Harold took the train to Rapid City and in the evening the train east to Chicago, watching the land progressively turn greener and with trees. Chapter includes a record of the birds and plants seen in the Badlands' area in the summer of 1920.

Keywords

Chalcedony · Scenic, south Dakota · Rapid City, South Dakota · Train travel

Monday, August 16

This morning we drove south from Scenic on the reservation road to a point about half a mile inside the Reservation fence about 8 miles south of Scenic. From here we ascended the green grass covered hill area east of Sheep Mountain and followed it for some distance till we reached the gap at the east end of Sheep Mountain. Here I took a three-section panorama of the east end of the mountain. We then went down to the pass which is about the level of the upper nodular layer and crossed over to the south side of the divide into a valley almost below Stony Pass. We found an exposure of what appeared to be the nodular layer here but very little material in it. I took three more pictures of various areas of pinnacles and then we prospected the canyon for some time. We also looked at an exposure of *Protoceras* sandstone which we had seen from the top of the mountain, but it was likewise barren of material. We noted here in one place that the intrusion of the silica which formed the chalcedony veins had impregnated and hardened the clays adjacent to the intrusion, and that the chalcedony did not easily separate from the hardened clay. This is the only point at which we had noticed this structure. We crossed back over the green hill to Sheep Mountain after some time and had a few canyons to cross in this before we got back on the side toward Scenic. This hill appeared to be of sand dune origin. We returned to the car and prospected the area in its vicinity but found that all the beds there must be higher than the nodular layer, so we returned to camp at about 5:30. We were all right until we reached the edge of Hart Table when the engine went dead on the hill for the first time, and Mr. Sinclair had to back down and start up. He made numerous attempts at the hill, but it was about the eighth time that he succeeded in reaching the top. When we got back to camp, he found that one of his coils was entirely dead and consequently was running on only three cylinders.

Tuesday, August 17

This morning Mr. Sinclair walked into town with the bad coil and got a new one to replace it. I went down into the badlands near camp to get some more pictures and also to collect some samples of the brilliant colored clays of the lower *Titanotherium* beds (Fig. 16.1). I went first along Hart Table for about 2 or 3 miles to the point where Bowen's road goes down into Indian Creek valley, and then down into the Pierre Shale badlands for some distance. I secured a picture or two of the transition between the Pierre Shale and *Titanotherium* beds and collected 8 samples of the various strata of bright colored clays ranging from white to gray, green, yellow, and brown. With this load, I returned to camp, about a mile and a half or 2 miles distant. In the afternoon I helped Mr. Sinclair clean the carbon from the cylinder head, and he gave the car a pretty general overhauling in order to have it in good shape for the return trip. We did no more prospecting this day.

Wednesday, August 18

This morning Mr. Sinclair and I went south by Hynz's road for Sheep Mountain. We then left the car and went up the canyon to the top and to Stony Pass. Mrs. Hynz gave us three of four ears of green corn. Mr. Hynz and most of the people of this section were starting today for Interior as the big annual 3-day roundup began today. We went on to the pinnacles of Sheep Mountain and I secured several pictures from this region near Stony Pass of the canyons and Sheep Mountain (Fig. 16.2). We then prospected the Stony Pass section of the upper white volcanic ash layer (Fig. 16.3) and a part of the east side of the mountain. Mr. Sinclair found a skull of *Leptauchenia*, the type animal of the upper beds here in a very hard matrix to extract from, but he got it out, and took it as it was the first one collected this year. We ate lunch under a pine tree which had been somewhat eaten by porcupines. We prospected some along the west side of the mountain at the heads of some of the canyons. We found nothing more collectable. There are many Chickadees and Chipping Sparrows, which had not been seen elsewhere, among the cedars and pines of Sheep Mountain [see listings at end of this chapter of birds (Sect. 16.1) and plants (Sect. 16.2) observed in this summer of 1920]. I got to a position where I could see the place where I had climbed up the mountain and where I had gone around the heads of side canyons to the south part of the mountain.

Fig. 16.1   View of the edge of Hart Table about a mile south of camp, showing the upper *Titanotherium* clays, the turtle *Oreodon* red layer, and the surface alluvium of the table. (Autochrome photograph taken 15 August 1920)

Fig. 16.2 View from the south end of Sheep Mountain (main part) of the great canyon of Indian Creek which has almost cut the mountain in two. The vertical walls are as high as 300 feet of more here, and it is practically impossible to ascend the mountain at this point. (Autochrome photograph taken 18 August 1920)

We also made some more observations of the creek valleys and badland areas to be seen from here. We then started back to camp at about 3 o'clock and stopped to have a little farewell visit with Mrs. Hynz, then back to the car at the base of the canyon.

This was our last day of collecting of the summer, and as a symbol of this we knocked the heads of the picks off their handles. We had collected, in a period of time from the 29th of June to the 18th of August, 94 numbered specimens of which 60 or 61 were complete skulls many of them with lower jaws and one a complete skeleton, three were feet, nine were perfect turtles and the rest were mostly lower jaws or sets of teeth [see **Appendix** for 1920 collected skulls and specimens now on display on-line through Yale Peabody Museum (YPM) vertebrate paleontology website]. We had also made some valuable stratigraphic observations and had learned that algae had a rather important part in the origin of some of the badland formations. I had also secured a good number of pictures which, if successful, will prove of considerable value. We returned to Scenic for our last night on Hart Table.

Thursday, August 19
This morning we picked up things in general, discarded some things not worth taking. I took a picture of camp (Fig. 16.4) and shortly after we took the tents down and packed them in the car. Mrs. Sinclair went into town with Mr. Sinclair, and I stayed with the rest of the stuff till his return. We then loaded everything remaining into the car and left Hart Table for good after being on it for 5 weeks and 4 days (Fig. 16.5). He already had the tent pitched in the schoolyard when we arrived, and we had lunch in it. We then made two more boxes in Mr. Bump's back yard in which we packed all the remaining specimens and little of the collecting material. We shipped these off. Scenic is almost deserted today as it was the second day of the big roundup. After supper, Mr. Sinclair and I looked at the low buttes just south of town and found a heavy cherty layer there which we had not observed elsewhere. On examination later in the labora-

Fig. 16.3 White ash layer exposed in Stony Pass between the northern and southern sections of Sheep Mountain. This is a peculiar apparently nodular material, made up of almost entirely volcanic ash, and a very poor matrix for fossils. It was in this section where we found the skull of a peccary. Faint colors of green, yellow, blue, and other shades are occasionally seen in this material, the cause of which we did not determine. (Autochrome photograph taken 18 August 1920)

Fig. 16.4 Our camp on Hart Table just before we broke camp and Professor and Mrs. Sinclair. Left, box tent of Mr. [William] and Mrs. [Delia] Sinclair. Right, my shelter tent, behind which is the car. Most of our paraphernalia can be seen in the background packed ready for transportation to town. (Photograph taken 19 August 1920)

tory this has turned out to be largely a partly silicified ostracod limestone. This was my last night in a sleeping bag.

Friday, August 20

This morning we packed up my tent and a number of other things in a box which we made, and

Fig. 16.5 The canyon nearest camp, showing the spring at the base, where we procured our water during our 6 weeks stay at Hart Table, and the Taylor ranch house. (Autochrome photograph taken 19 August 1920)

also included a good deal of camp equipment which would be unnecessary for the return trip. We also shipped this off. There were very few people left in town today as it was the last day of the roundup and a big day. Everybody here seemed to be talking about the bull dogging, steer throwing, etc., and lots of tourists were going through. I expected to leave today, but the bank was closed so no money was available for my railroad fare, and it was necessary for me to stay over another day. Miss Leona Malloy came to make a short call on us this afternoon before she went back to her sister's place in the Badlands. In the afternoon, we had nothing to do, as everything was packed up. I packed my bedroll in my duffle bag and everything that would not go in this, in my bag. I had to stay at the hotel this night as my bed and tent were both not available. It seemed very stuffy after sleeping all summer in the open, and I could not sleep very much. Just before bedtime, a special car on a freight train brought back most of the people of Scenic from the roundup.

Saturday, August 21

This morning I had my last meal in camp with the Sinclairs. They got everything packed in the car by about 9 o'clock. We then went over to the main street and Mr. Sinclair met Mr. Loomis, professor at Amherst, an old friend of Mr. Sinclair's who has been collecting in badlands of Wyoming, Nebraska, and South Dakota with five student assistants. The Sinclairs pulled out about 9:15 for Interior and points east. I had about 4 or 5 hours to wait but had secured the money for transportation all right. Time passed very slowly as I had nothing to do, but I wrote two letters and said good bye to the postmaster, the boys at the Rochdale store, Mr. Bump and his son, Mr. and Mrs. Taylor and Guy Taylor, Dannie Malloy and other friends and acquaintances of the summer. I had lunch at Mrs. Taylors. I left Scenic at about 2:30 in the afternoon, arriving at Rapid City at about 4:00. I bought my ticket to Chicago and checked my duffle bag, then I went up Hangman's Hill and took a few pictures of the hills and the town (Fig. 16.6). I then had supper in town and

Fig. 16.6 View northeast from Hangman's Hill, Rapid City, showing the northern section of the city, and to the left, the hogback ridge of Dakota Sandstone, marking the first range of the Black hills. In the upper center is a large lumber yard, as Rapid City is an important lumber center. Note the flatness of the plain to the right. A treeless skyline of about 100 miles could be seen from here. (Photograph taken 21 August 1920)

took the train a little before dark. I had an upper berth.

Sunday, August 22

I did not sleep very much as the roadbed was very bumpy. I woke up near Long Pine, Nebraska and had breakfast there. Nebraska seems much greener than South Dakota with many trees and lots of farms. Corn seems to be good and large. I had lunch at South Norfolk, and supper at Omaha. I changed trains at Omaha and got an express train for Chicago. We crossed the Missouri between Omaha and Council Bluffs, Iowa, and I saw more trees in the next few miles than I had all summer.

Monday, August 23

I woke up near DeKalb, Ill., and arrived in Chicago at about 9 o'clock. I went home immediately, getting there at about 9:30, 2 months and 3 days after I had left.

    THE END

16.1 Record of Birds Seen in Badland Area, 1920

1. **Teal**? One seen in pond south of Scenic. Species could not be determined.
2. **Great Blue Heron.** One seen flying over Badlands at sunset.
3. **Wilson Phalarope.** A flock of 10 or so in a small marsh near Scenic.
4. **Upland Plover.** Two seen on Hart Table, on August 2nd.
5. **Killdeer.** Common in wet places and through prairie
6. **Turkey Buzzard.** Common in Badlands. Nests around Sheep Mountain.
7. **Red Shouldered Hawk.** Common in Badlands.
8. **Marsh Hawk.** Common in Badlands. Other hawks also seen, which could not be identified.
9. **Sparrow Hawk.** Rather common throughout area.

10. **Burrowing Owl.** On fence posts along roads and in Prairie Dog towns.
11. **Prairie Chicken.** One seen in valley of White River and two on Hart Table.
12. **Quail?** Flock on Hart Table on Morning of August 16th. Species unknown.
13. **Red Headed Woodpecker.** Three or four seen in valley of Indian Creek,
14. **Flicker.** One on Brown's farm near Interior, several in valley of Indian Creek and on Sheep Mountain.
15. **Night Hawk.** Quite common.
16. **Mourning Dove.** Very common in Badlands.
17. **Arkansas Kingbird.** Quite common on prairie and in creek bottoms.
18. **Kingbird.** Two seen along White River at Interior.
19. **Say's Phoebe.** Rather common in Badlands.
20. **Desert Horned Lark.** Very common on prairie and along roads. Commonest bird of the area.
21. **Crow.** One seen on Hart Table.
22. **Magpie.** Two seen from train near Imlay. Common in valley of Indian Creek.
23. **Western Meadowlark.** Abundant in Prairies and sings most of the time.
24. **Brewer's Blackbird.** One seen south of Scenic and one nested in the canyon of Taylor's Spring, near camp.
25. **Grackle.** Two seen in the town of Scenic.
26. **Red Winged Blackbird.** Rather common in small marshes.
27. **Yellow Headed Blackbird.** Two seen on street in Interior.
28. **White Throated Swift.** Rather common in higher cliffs, Sheep Mountain, great Wall, etc. Has very rapid flight.
29. **Western Chipping Sparrow.** On Sheep Mountain among pines and cedars.
30. **Western Grasshopper Sparrow.** Common on Hart Table in Prairie Grass.
31. **Lark Sparrow.** Rather common in Badlands and in Cedars at Interior.
32. **Towhee.** One or two seen in Cedar Pass, and in Sheep Mountain Cedars.
33. **Dickcissel.** One seen east of Interior on Washington Highway.
34. **Lark Bunting.** Rather common in prairie and abundant on Hart Table.
35. **Sparrow?** White tail feathers. May be **Baird Sparrow** or **Smith's Longspur.** Seen in oat stubble field August 16, in a small flock.
36. **White Rumped Shrike.** Common along Railroad and telephone lines.
37. **Cliff Swallow.** Commonest bird in towns of Scenic and Interior. Build colonies of nests in inaccessible cliffs as in Cliff Swallow Canyon.
38. **Barn Swallow.** One seen on Hart Table, August 14th.
39. **Brown Thrasher.** Nested in canyon of Taylor's spring near camp.
40. **Rock Wren.** Very common in Badlands and buttes. Sings all the time.
41. **Chickadee.** Rather common on Sheep Mountain.
42. **Mountain Bluebird.** Common in Badlands. Hovers in the air.
43. **Cowbird.** A small flock seen at Hart Table.

16.2 Partial List of Plants Identified on 1920 Expedition

[Common and scientific names in brackets are modern names after Johnson and Larson, 1999 and 2007]

1. *Pinus scopulorum.* **Western Yellow Pine.** Common on Sheep Mountain.
   [Ponderosa pine, *Pinus ponderosa*]
2. *Juniperus ecopulorum* [sic]. **Red Cedar.** Common in badland canyons.
   [Rocky Mountain juniper, *Juniperus scopulorum*]
3. *Bouteloua oligostachya.* **Grama Grass.** Abundant on all table lands.
   [Blue grama, *Bouteloua gracilis*, and sideoats grama, *Bouteloua curtipendula*]
4. *Tradescantila occidentalis.* **Spider wort.** Common in White River valley near Interior.
   [also bracted spiderwort, *Tradescantia bracteata*, which is locally more common]

5. *Calicortis* sp. **Maraposa Lily** or **Gumbo Lily**. Abundant on tables in June.

    [Sego lily, *Calochortus nuttallii;* gumbo lily now refers to the gumbo evening primrose]

6. *Yucca glauca.* **Yucca.** Common on edges of tables and in badlands.

7. *Populus sargentii.* **Western Cottonwood.** Common in badland valleys.

    [Plains cottonwood, *Populus deltoides*]

8. *Celtis occidentalis.* **Hackberry.** A few seen about 2 miles east of Sheep Mountain.

9. *Sarccbatus vermiculatus.* **Greaseweed.** Common in Badlands. [Greasewood]

10. *Salsola pastifer.* **Russian Thistle.** Common on tables and edges of tables.

    [also known as tumbleweed, *Salsola tragus*, an Asiatic invasive plant]

11. *Argemone intermedia.* **Prickly Poppy.** Common on Hart Table.

    [Crested prickleypoppy, *Argemone polyanthmemos*]

12. *Ribes americanum* or *floridum.* **Black Currant**. Common in moister sections, as on the banks of the White River. [Probably a variant of the golden currant, *R. aureum*]

13. *Ribes aureum.* **Red Currant.** Common in badland canyons where cedars grow.

    [Golden currant, the common currant of the Great Plains]

14. *Rosa arkansana.* **Wild Rose.** Common in prairie. [Prairie rose]

15. *Prunus besseyi.* **Sand Cherry.** A few seen on edge of Hart Table.

    [Chokecherry, *Prunus virginiana*]

16. *Astragalus canadansis.* **Vetch.** Common. [Probably includes many species of *Astragalus*, including milkvetch and poisonvetch]

17. *Aragallus lambertii.* **Loco.** Common in badland slopes. [Locoweed or Lambert's crazyweed]

18. *Psoralea digitata.* **"Wild Alfalfa."** Common on Hart Table. Many other legumes which are not known specifically. [Alfalfa, *Medicargo falcata* crossed with. *M. sativa*]

19. *Petalostemon olighphyllus.* **Prairie clover.** Common on Hart Table, especially on the edges. [Includes the white and purple prairie clover, *Dalea candida* and *D. purpurea*]

20. *Euphorbea marginata.* **Snow on the Mountain**. Common in alkaline badland flats.

21. *Rhus trilobata.* **Skunkbush.** Common in cedar zones

22. *Acer negundo.* **Box elder**. A few near Taylor's Ranch house.

23. *Mentzalia decapetala.* **Loasa.** Night blooming. Very showy, common on Hart Table.

    [Tenpetal blazingstar]

24. *Opuntia polyacantha.* **Cactus (Prickly Pear).** Abundant in Badlands. [Plains pricklypear] (Fig. 16.7)

25. *Sheperdia argentea.* **Buffalo Berry**. Common in Indian Creek and other Badland areas.

    [Silver buffaloberry]

26. *Anogra albicaulis*, **White Evening Primrose.** One of the most common plants on the hard Badlands. [Gumbo evening primrose, gumbo lily, *Oenothera caespitosa*]

27. *Fraxinus lanceolate.* **Green ash.** Found about a mile south of camp at base of Hart Table.

    [*Fraxinus pennsylvanica*]

28. *Apocynum cannabinum.* **Dogbane**. Common in cedar zone, Cedar Pass and Sheep Mountain.

    [Indian hemp]

29. *Accrates angustifolia.* **Green Milkweed.** On Hart Table. [*Asclepias viridifolia*]

30. *Asclepias speciosa.* **Common Milkweed.** Abundant on edge of Hart Table.

    [Showy milkweed]

31. *Ipomoea leptophylla.* **Bush Morning Glory.** Common on Hart Table and Sheep Mountain.

32. *Solanum rostratum.* **Buffalo Bur**, Thistletomato. Unpleasant odor. Common in valley of Indian Creek in places.

33. *Plantago purshii.* **Prairie Plantain.** Abundant on the unbroken prairie.

    [Indianwheat or wooly plantain, *Plantago patagonica*]

Fig. 16.7   [Prickly Pear] Cactus – Corral Draw. (Autochrome photograph taken in 1921)

34. *Liatris punctate*. **Smaller blazing star**. Abundant on Hart Table.

    [Gayfeather, *Liatris puncata* is the dotted gayfeather]

35. *Chrysothamnus graveolens*. **Rabbit brush.** Very common in the Badlands.

    [Rubber rabbitbrush, *Chrysothamnus nauseosus*]

36. *Rapibida columnaris*. **Cone flower.** Common on Hart Table.

    [Prairie coneflower, *Ratibida columifera*]

37. *Ratibida columnaris pulcherrima*. **Red cone flower.** A few near Interior.

    [Probably a variation of the Prairie coneflower]

38. *Artemisia*. **Eastern Sage**. Common on tables and edges.

    [Several species of *Artemisia* occur in the Badlands, none now called eastern sage].

39. *Artemisia longifolia*. **Sage Brush**. Abundant in the Badlands.

    [Silver sagebrush, *Artemisia cana*].

40. *Carduus altissimum*. **Common thistle**. Common. [There are several species of thistle in Badlands, and only one species of *Carduus*, musk thistle, *C. nutans*]

41. *Tragopogon parvifolius*. **Oyster plant**. Abundant on Hart Table. [Goatsbeard, *Tragopogon dubius*; the oysterplant, *T. porrifolius*, is rare in South Dakota; both are Eurasian plants].

42. *Bidens tricosperma tenuiloba*. **Fennel** or **Fetid Marigold**. Common in Indian Creek.

    [Fetid marigold, *Dyssodia papposa*]

43. *Malvastrum cocoineum*. **False Mallow**. Red. Common in prairies.

    [Scarlet globemallow, *Sphaeralcea coccinea*]

44. *Helianthus peticlaris*. **Low Sunflower.** One of the most abundant badland plants.

    [Plains sunflower]

45. *Delphinium* sp. **Tall White Larkspur**. Common on table lands.

    [Prairie larkspur, *Delphinium virescens*]

46. **Blue foxglove**. Most brilliant flower of the summer. On edge of Hart Table.

    [Smooth beardtongue, *Penstemon glaber*]

47. **Smaller lavender foxglove**. Also, quite common on Hart Table.

    [Narrowleaf beardtongue, *Penstemon augustifolius*]

Many others not identified.

References

Johnson JR Larson GE (1999) Grassland plants of South Dakota and the northern Great Plains (revised). South Dakota State University College of Agriculture and Biological Sciences B566, 288 p

Johnson JR Larson GE (2007) Plants of the Black Hills and Bear Lodge Mountains, 2nd edn. South Dakota State University College of Agriculture and Biological Sciences B732, 608 p

Part III

Excerpts from the 1921 and 1922 Expeditions

Part III (Chaps. 17 and 18)

Abstract   These two chapters are excerpts of Wanless' 1921 and 1922 field seasons back in the Badlands where he was both collecting fossil skulls and also doing the necessary field geology for his Ph.D. degree at Princeton. The 1921 season included Wanless purchasing a 1916 Ford Model-T, learning to drive it, and driving from New Jersey to South Dakota and back. The 1922 season was with his mentor Professor Sinclair. The work was mostly science, and Wanless was constantly addressing the meaning of geologic observations for reconstructing the ancient environments represented in the sediment sequences. In addition, the letters provide important additional insight into the nature of the people and the landscape of the Badlands.

Tackling the West in the *Rachael Jane*: 1921 Expedition

17

Abstract

Wanless' 1921 field seasons back in the Badlands was both collecting vertebrate fossils and doing the necessary field geology for his sedimentology and stratigraphy Ph.D. degree at Princeton. The 1921 season included Wanless purchasing a 1916 Ford Model-T, learning to drive it, and driving from New Jersey to South Dakota and back with field assistants. His naturalist mother, Rhoda, also came west by train to join them. Harold describes and illustrates numerous sandstone-filled channels in the White River rocks in the headwaters of Big Sand Draw and illustrates sand crystals on Rattlesnake Butte. They also take side trips to the Black Hills. In addition, the letters provide important additional insight into the nature of the people and the landscape of the Badlands and the travelling in a Model-T. Trip expenses are included.

Keywords

Model-T Ford · Rhoda Wanless · Sandstone channels · Sand crystals · Big Sand Draw · Black Hills

I asked the old Black Man, "What is that bird that sings so well?" He answered: "That is the Rachel-Jane."
"Hasn't it another name, lark, or thrush, or the like?" "No. Jus' Rachel-Jane."
Modified from *The Santa Fe Trail--A Humoresque* by Vachel Lindsay, July 1914

In the fall of 1920, Harold Wanless, as a part of his Master's program, worked on his two-volume geologic studies on the Rosendale Quadrangle in New York state, about 90 miles north of New York City. He also printed the dozens of photographs that he had taken the previous summer in the Badlands. William Sinclair prepared and catalogued the 100 fossil specimens collected in 1920. He also worked on three manuscripts on the geology of the "red layer," the entelodonts, and a sabretooth cat collected in 1920. All three of these papers were published in the *Proceedings of the American Philosophical Society* in 1921 (Sinclair 1921a, b, c). The "red layer" paper included photographs taken by Wanless.

© The Author(s), under exclusive license to Springer Nature Switzerland AG 2023
H. R. Wanless, E. Evanoff, *The Diaries of a Bonedigger*,
https://doi.org/10.1007/978-3-031-25118-4_17

Princeton, Nov. 22, 1920 (**excerpted from**)

Dear Mother,

. . . . I had dinner down at Sinclair's Sunday noon and took the prints along. They were very much interested, and Mr. Sinclair made the suggestions I mentioned. . . . . We had many reminiscences of the summer. By the way, a great misfortune has come. It appears that in all probability the big pig skull is different from any hitherto collected, and Mr. Sinclair declares that unless he finds it like some other known form, he is going to name it after me, Entelodon wanlessi. The fossil bird's egg which I found is now down in the museum, and another of the smaller Entelodon skulls is prepared but not placed yet as he is keeping it handy for reference in working on the others. It seems to be Entelodon clavis, a described form.

In the spring of 1921, Harold finished his Master's degree submitting a two-volume manuscript on the geology of the Rosendale Quadrangle. He also made time to compile, type, and illustrate his 1920 field journal (the narrative and plates of Chaps. 3, 4, 5, 6, 7, 8, 9, 10, 11, 12, 13, 14, 15, and 16 in this book), which he completed in March 1921.

In April, Harold decided that he needed a car for his field work, and he bought a used 1916 Model T Ford touring car for $250. He financed it by drumming up funds from four of his friends (including his future brother in law) with the agreement that he would pay them back for the full amount when he left for the field in June (Fig. 17.1). Harold named his car the "Rachel Jane" after hearing the poem "*The Santa Fe Trail-A Humoresque*" read by the poet Vachel Lindsay at a recital in Princeton[1], [2]. In the poem, Rachel Jane was a name of a small songbird

beside a road filled with noisy cars heading west (see below). Since Sinclair had not allowed Harold to drive his car in the 1920 Expedition, he had to learn to drive in May and early June 1921[60]. In addition, Wanless made some modifications to the car for the trip west.[3]

Princeton, May 12, 1921.

Dear Mother,

I got your good letter, enclosing the check, and was very glad to hear about your Lakewood trip. I have taken out the Public Liability Insurance on the car, and we have already gotten the new tire. We got a United States tire, which is supposed to be quite a good kind, although less in price than some of the Standard tires. A week ago yesterday, Walter Dew persuaded me to go down to Trenton to take the state driver's examination, even though it was only three days since I had first tried driving. I was quite worried as to whether or not I would pass it, but I had little difficulty with either the practical demonstration or the written test on the Motor Vehicle Law and Traffic Regulations of the state of New Jersey. . . . .

I did my first city driving yesterday, as I drove for a couple of miles in the city of Trenton. It is much harder than country driving, as one must keep on the alert every minute for street cars, other automobiles, school children, and pedestrians in general. I have gotten quite confident in country driving but driving where there is a good deal of traffic gets me very nervous. I have driven about 60 miles so far, I think and have come no nearer an accident than stalling the engine on a car track in the town called Hopewell, 9 miles from Princeton. . . . .

---

[1] A poem most worth hearing: https://www.youtube.com/watch?v=Y2gLY5Jt-xo

[2] Wanless took his tests to qualify for a New Jersey driver's license on the fourth day after his first drive. He passed.

[3] "Another expense which I will have to put into the car will be to fit the front wheels with demountable rims, which will make the front and back tires the same size. At present, the front tires are 30 × 3, and the back ones are

Fig. 17.1 The "Rachel Jane," the 1916 Model T Ford touring car in which Harold made a trip from Princeton, New Jersey, to the Badlands of South Dakota in 1921. Harold Wanless, with the help of his four friends, bought the car in late April 1921. (Pictured are the original 5 student purchasers: Arthur Horton, Maury Rogers, Mac Harper, Ralph Beebe, and Harold Wanless (on right))

I think I told you about the name of the car in my last letter, but if I did not, I will tell you now. Vachel Lindsay, a modern poet who has a very particular style of poetry, which aims to portray moods of various sorts, gave a reading of poems here about two or three weeks ago. One of the poems had to do with the streams of automobiles going westward straight mile after mile along the Santa Fe Trail with horns of all sorts from morning to night, and how above all the human noise could be heard the song of a little bird in a hedge along the road singing "Sweet, sweet, sweet." He said he asked someone what the name of the bird was, and they said "Rachael Jane," and so he has called it that ever since. So, we therefore decided unanimously to name our new car the Rachael Jane, and it is already widely known by this name. . . . .

I heard the first of a series of 5 lectures which are being given here this week by Professor Einstein, the author of the famous theory of relativity. It was in German and not very clearly enunciated, so that it was very difficult to understand. A
twenty minute resume of it in English was given at the close by Professor Adams of the Department of Physics. I understood this, but I decided that it was not worthwhile to hear the rest of the series of lectures this week. Princeton is the only place in the country where Einstein is going to explain the theory of relativity.

[Harold was not alone. Over 500 people had tickets to the first lecture; less than 50 attended the final lecture.]

On June 26 Harold left Princeton in the Rachel Jane with two friends who would be his field assistants in the Badlands during the summer of 1921. These were Walter A. Dew and George A. Wiggan. Rhoda Wanless, Harold's mother, also assisted (Fig. 17.2) but arrived in Scenic by train. They camped on Hart Table (Fig. 17.3) and

30 × 3 1/2, and fitted with demountable rims. The cost of the rims will be $10.00 for the pair, and the tires will cost about $18.00 a piece with inner tubes. This will make a total of about $46.00" (from letter to Rhoda Wanless from Princeton, 2 May 1921).

Fig. 17.2 Harold Wanless' photographs of his field crew during the summer of 1921: (**a**) Walter A. Dew (Princeton class of 1921), (**b**) George A. Wiggan (Princeton class of 1923), and (**c**) Rhoda E. Wanless, Harold's mother, on Sheep Mountain

Fig. 17.3 The party of the 1921 Expedition assisting Wanless at the permanent camp on Hart Table: Walter E. Dew, George Wiggan, and Rhoda E. Wanless. (The photograph also shows the car "Rachael Jane")

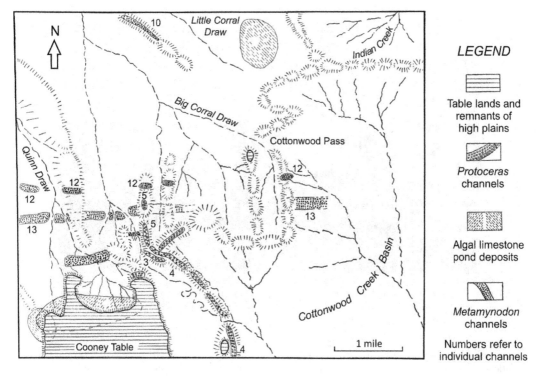

Fig. 17.4 Map of the tables and Badlands of the Corral Draw, Quinn Draw, and Cottonwood Creek Basin southwest of Scenic. Mapped are the contrasting lateral trends of the *Metamynodon* channels (solid circles) and *Protoceras* channels (open circles), and the areas of algal limestones are also noted. Map lettering has been redrafted for clarity

worked for most of the summer concentrating on an area west of Sheep Mountain Table, approximately 12 miles to the south and southwest of their camp on the Taylor Ranch. This area included the headwaters of Little and Big Corral draws, Quinn Draw, and Battle Creek Draw[4] (Figs. 7.1 and 17.4). They also camped for a time at the head of Corral Draw near access to many exposures of fossil river channel sandstones (Fig. 17.5). Harold made descriptions of the geology and all four collected a total of 140 vertebrate fossils for the Princeton Natural History Museum. They also made two trips to the Black Hills, the first a 4-day tour of the Hills in early August, and the second on August 31 and September 1 to collect minerals from the pegmatite mines. George Wiggan had to leave the crew on August 3 to attend to his duties as an Army National Guard reservist. Rhoda Wanless left Scenic for her Chicago home by train about September 3.

Harold and Walter remained, boxed up and sent the fossil specimens to Princeton, and packed up the camp, leaving Scenic on September 10.

Excerpt from the 1921 trip west by Harold Rollin Wanless (1921a):

Last April the writer learned that he was to conduct an expedition to the South Dakota Badlands to continue the geological work which had been begun the summer before with Professor Sinclair.
The plan was to work the famous section of Badlands in Corral and Quin Draws, from which John Bell Hatcher, collecting for Princeton in the 1890s, brought out some of the finest fossilized mammals ever found. As far as we know, no geologists had since collected in this area and we felt sure that the erosion of twenty-five years would have washed out many things of interest. The Corral Draw district is , however, a difficult section to work, as it is situated twenty-five miles from the railroad and is uninhabited save by a few cattle and horses, which find sparse grazing in its basin. Furthermore, there is no good water within fifteen miles.
"It was decided that the best way to get a party to the Badlands would be to drive out. Consequently,

---

[4]Now known as Nevis Draw.

Fig. 17.5 The "Camp Channel" at the head of Corral Draw, showing the excellent development of crossbedding in the *Metamynodon* channel sandstones here. Walter Dew and Mrs. Rhoda Wanless are in the foreground. (Black and white photograph taken in 1921, hand colored by Rhoda Wanless)

a second-hand Ford was purchased and fixed up a little to stand the rough wear of the Badlands. We were given the positive assurance from most of the graduate students that the car would never be able to cross the Appalachian Mountains. However, on June 26th the writer, accompanied by Walter A. Dew '21 and George A. Wiggan '23 started on the long journey and followed the Lincoln highway westward. After the mountains had been crossed on our second day out, there was a long ride across the flat agricultural states of the Mississippi Valley and over the very poor roads of Iowa till the Missouri River was reached on the 4th of July. Upon crossing the Missouri River, we entered the free west, where one can stand on a high hill and look off for miles without seeing a tree or fence-post or a house. When we were within one hundred miles of our destination, we made the mistake of camping for the night six miles out of one town and ten miles from the next. The result was we were unable to proceed and had to wade back through six miles of mud for breakfast. However, two days later, our car, the "Rachael Jane," drove into Scenic, South Dakota, our base of supplies for the summer.

"We had fully expected to put up the car on our arrival and use team and wagon all summer, haul-ing our water fifteen miles or more, and so we set out on horse-back to pick a site. We learned of a small spring of water in the Badlands and located there. We camped here for a week, hiking twenty or twenty-five miles a day, with the total result of two collections for the week. During this time, we dis-covered a dim Indian trail up into Corral Draw and tried to get in with the car. The first attempt

resulted in a sheared hub and a broken front frame, but, after a while, the "Rachael Jane" learned the ways of the Badlands and took us almost daily into Corral Draw country (Figs. 17.4 and 17.5), allow-ing us the luxury of a camp near good water and civilization. The result was that next week regis-tered for us eighteen collections of quite good quality. From then on, our success was uniform, and collections continued to come in rapidly.

"The collections consisted of the fossilized remains of animals of the Tertiary Period, which had lived in this Dakota country many million years ago. They are of special interest as they include the ancestors of many important domestic and wild animals of the present day. Among the summer's collections were rhinoceroses of three kinds, giant pigs, three-toes horses, ancestral camels, sabret-oothed tigers, dogs, peccaries, turtles, etc. The size variation ranges from Palaeolagus, the ancestral rabbit, which is a smaller than a modern rat, to Titanotherium, the huge horned rhinoceroslike beast, which was as big as the modern elephant. Among our summer's collections were a skull and the lower jaws of two different species of Colodon, a small tapir-like animal heretofore known only from fragments. The collections of Hyracodon, a horse-like rhinoceros, have already enabled Professor Sinclair to revise the species of this animal."

In addition to collecting specimens, much of the field work that Harold used for his doctoral dis-sertation was done in the summer of 1921. This consisted principally of measuring and describ-

Fig. 17.6 View from the top of Sheep Mountain, of a canyon in the *Oreodon* beds. Pinnacles of the *Leptauchenia* ash beds are visible to the right. This canyon shows a pale blue color, due to the pieces of chalcedony weathered out from chalcedony veins on its sides. The grass colored hill in the distance is a line of sand hills on which Mr. Arnold's ranch is located (see Fig. 7.1). (Autochrome photograph taken in 1921)

ing vertical sections of the sedimentary sequences (stratigraphy) in the Sheep Mountain area (Fig. 17.6).

In the process of collecting vertebrate fossils and in describing and measuring the vertical stratigraphic sequences in the western White River Badlands, Wanless began observing and contemplating the meaning of a variety of characteristics of the sediments and the ancient environments in which they accumulated.

In the Corral Draw area, including near their camp there, sandstone deposits of ancient river channels were a focus of interest (Fig. 17.7). He noted that the channel patterns and orientations would change upwards through the sequence. For example, the channels within the *Titanotherium* beds had "somewhat different courses" than the *Metamynodon* channels 50 feet above in the lower Oreodon beds directly above (Wanless 1921b, Note VII).

Within the *Metamynodon* channels, he recorded that the channel sand width would expand and diminish up through the channel's 27 ½ feet thickness: "The lower levels of the channel seem to extend further laterally than the upper

levels, as if they marked a more fluviatile [stronger river discharge] epoch (Fig. 17.8). At three points the sands send out lateral fingers into the clays, between which the clays extend further in toward the center of the channel, as if the three points represented spells when the stream bed was expanded more widely due to periods of high precipitation" (Wanless 1921b, Note VIII). This is the important beginning of considering climate fluctuations in his reconstructive interpretations of ancient environments.

Wanless was also making observations towards understanding the meaning of the limestone deposits within the terrestrial environments of the White River Badland's beds. He described both limestone crust layers and beds with individual calcified algal balls. In little Corral Draw, he describes a: "Heavy bed of algal limestone about 2 feet thick, in the lower Oreodon beds occurs 27 ½ feet above the base of the *Oreodon* beds. This bed seems to be in a general northwest southeast direction for a mile of two, and has a rather lens like shape, as if it represented deposition in a pond" (Wanless 1921b, Note VI). As limestone needs supersaturated waters, replen-

Fig. 17.7 Capping channel sandstone in Corral Draw, about 1/4 mile east of Camp. (Autochrome photograph taken in 1921)

Fig. 17.8 Camp channel section, near the head of Corral Draw. Beds of sandstone extending out into the clay at three levels to the left are visible, which may represent cycles of moister conditions [producing stronger flood discharge]

ishment of water, a lack of being overwhelmed by silt and clay, and time to form, he is striving to envision an area and climate alongside rivers where these conditions can be met.

In one case a "heavy [meaning solidly lith-ified] white limestone crust" is capping an ero-sional surface at the top of the *Titanotherium* beds (Wanless 1921b, Note IX). This he realizes

is formed during a prolonged time of exposure, possibly from groundwaters drawn up to the sur-face and evaporated.

In other cases, he describes the limestone not as crusts but as a great number of "concentric[ly laminated] algal balls" (Wanless 1921b, Note XIII), recognizing a formation by exposure to supersaturated water and probably agitation.

Fig. 17.9 A mass of sand calcite crystals, photographed in place at Rattlesnake Butte, Pine Ridge Reservation, South Dakota. This shows the excellent development of the scalenohedral form

One of the features documented and later published on were "sand crystals." These are quartz sands in which large single crystals of calcite grew within the pore space between the quartz sand grains (Fig. 17.9).

The other aspects of the 1921 summer in the White River Badlands and the adventures with Rachael Jane are perhaps best shared through a series of letters to Harold's future wife and his mother. Especially enjoy the vibrance of Scenic in its 1921 heyday.

Letter to Grace Rogers (Harold's future wife) in Ashbourne, Pennsylvania

Scenic, Pennington County, South Dakota,
August 13, 1921

Dear Grace:

I received your card and was glad to know you are getting a sunburn at least. I passed that stage at least a month ago and am now hardly distinguishable from a Sioux Indian in shade of color.

Rachael Jane behaved excellently on the trip out, making it in seven days, in spite of the fact that we struck rain on five of these days, and had to lay over a whole day in Vivian, S. D. when she got stuck in the mud on a steep hill. She has caused quite a lot of difficulty here, having broken the front spring twice, the front frame once, and the driving shaft housing and ring gear once.

We have been here now for about five weeks and have made 64 collections which is coming very nicely, as our last year's total was 96. We have gotten a big variety of things, all the way from an ancestral rabbit skull about on inch long to the skull of a Titanotherium animal 13 feet long and 8 feet high.

We have not used a team and wagon as we expected to but have camped on Hart Table in our old campsite and been using the car to get into the rough badlands country we are working. It is 25 miles from our campsite to Corral Draw by the route we use, and we have made the trip 8 or ten times. The roads are terrible most of the way but passable, and we thought it better than $4.00 a day for a team and wagon all summer and having to haul water ten or 15 miles.

George Wiggan, the fourth member of our party, had to leave us a week ago last Wednesday to go to a reserve officer's camp in Maryland. Just before he left, we took a four-day trip through the Black Hills. I rained most of the time we were there, but we had a very interesting time anyway. The country is about the finest I have ever been in. We stayed overnight the first night at Sylvan Lake, up near Harney's Peak. There are vertical pinnacles of rock hundreds of feet high in all directions from the lake and great tall spires over 100 feet high around the lake and nearby canyons.

The next day, we went to Lead and there the Homestake gold mine, the largest gold mine in the country.

We also drove the Spearfish Canyon, a beautiful canyon about 40 miles long with dense woods at the base and cliffs 1,000 feet or more high on the sides. Here we actually went swimming in an icy mountain brook, and then just afterward that had our first blowout since Pittsburg [Pennsylvania]. It was the tire we bought in Trenton the night you were along, and it made me think of our blowout there. We then drove Rachael up to the top of Mt. Roosevelt about 1500 feet above Deadwood and a seven mile pull up from the town over a rocky road. From the top, we could see Wyoming, Montana, and North Dakota. We drove back to Rapid City and left George, and then to Scenic and started work again.

They call this an arid climate, but we have had at least ten or twelve rains here in the past two weeks, and it had rained now every day for three days. Yesterday, Hart Table had a hail storm for about an hour and mother was alone in the tent during it. Both last night and the night before, there were full arches of rainbows, and the night before that we could see three different rainbows. It has cleared tonight, and we hope it will stay clear for a while.

You may be interested in the way Scenic is going to open its new congregational church tomorrow. There is going to be a sermon at eleven, then a community dinner at which everyone is going to bring his own lunch. Then, in the afternoon, there is going to be a baseball game and a bucking steer contest, and a wrestling match, and then another sermon in the evening.

Scenic is boring for oil now one mile north of town, and there is much interest in it among the inhabitants. The have formed a local company and Mr. Kennedy, the hotel keeper, has been after me every chance he has had to get me to buy some stock in it. They struck a show of oil and gas explosions only 80 feet down and feel very optimistic of their chances of striking oil. If they get a producing well, it will be the first in the State, and they will get a $50,000 State bonus. When they started drilling, they had a "spudding in" day, and all the ranchmen and farmers around for miles came in. They had a barbeque and served hot roast beef sandwiches and coffee free to everybody. Then at four, made speeches on the momentous occasion for Scenic and what oil would do for the town, and they had two ball games, then a prize bucking horse contest. I guess it was Scenic's biggest day.

The driller of the oil well located the top of a titanothere skull near the well and told us of it and we spent two days excavating it from its tough sandstone matrix. It is 30 inches long, a foot wide and a foot high and weighs over 100 pounds. It will be a fine museum specimen.

Well, I guess it's getting about bedtime (9 o'clock) as we are usually early risers here. Give my regards to your father and mother and all the rest.

Sincerely yours, Harold R. Wanless

Letter to his Mother as he is leaving the Badlands:

Scenic, Sept. 4, 1921

Dear Mother,

We got our section at Sheep Mountain measured all right, but it took till nearly six o'clock, so we got supper in town. The potatoes spoiled as we forgot them this morning.

We were glad you were not here last night as you would have almost frozen to death. There was a terrible north wind gale all night with the temperature not far from freezing. By far the coldest night of the summer to date. Tonight is also cold, but there is no strong wind.

This afternoon, we collected your Hydrocodon. The jaws are excellent, but no skull was present. Also shellacked my titanothere and collected Walt's, giving us 119 collections already written up.

By the way, I forgot to give you the address of the Electrical School from Wesley Taylor, 39-51 East Illinois Street. Do not forget to write Mrs. Taylor about it right away. I do not know Mr. Taylor's first name, so to avoid trouble perhaps you had better address the letter to Wesley Taylor, Caputa, South Dakota, unless you know Mr. Taylor's 1st name.

We wrapped specimens last night till 11:30 and I wrapped the Bear Creek specimen today, making about 200 packages wrapped since you left. We are now down to 4 sheets of Cosmopolitan magazine. We stopped at Taylor's a few moments last night and I apologized for your not coming down. Mrs. Taylor said she was so busy making bread she did not have time to come up. I am going to take Earl's picture on horseback tomorrow morning. We will also put a paste cloth on the titanothere, then get a good early start for Corral Draw.

I will enclose the receipt for the box and will get it from the agent tomorrow morning. I must be getting to bed, as it is getting both cold and late.

Lots of love, Harold
P.S. No trouble with Rachael yet. Mrs. Taylor's address is Mrs. Thomas H. Taylor, Cabuta, S. Dakota.

Letter written to Harold's mother after visiting her in Chicago on the way back east.

Princeton, Sept. 27, 1921

Dear Mother,

The Rachael Jane and its occupants arrived in Princeton yesterday morning at 1 a.m., thus completing the expedition. After we left Jim Benson's place, we had no trouble with her except, that in going over the mountains, we wore out the reverse band, and for a hundred miles or so were unable to back up the car. However, it has never been the policy of Rachael to go backwards anyway, so we moved ahead as usual and arrived in Shippensburg about 11:00 Saturday morning. We have new brake bands put in while we were there and left Shippensburg the next morning at about 9:00.

I received your letter and was very glad to hear from you but sorry you were feeling so lonesome.

We drove from Jim Benson's farm down to Salem on Thursday, arriving in Salem at about 4:00. We stopped and called at the Stratton's, and Walt showed me his scrap collection, which had been put on exhibition in the window of Mr. Stratton's store. There was also an article in the Salem paper telling of Walt's deeds and discoveries. We had supper with Walt's aunt and then started on, crossing the Ohio River 30 miles from Salem at East Liverpool. We reached this point about dark. We had very hilly country from here to Washington, Pa., and the roads were partly dirt roads, but we got along and arrived at Walter's aunt's home in Washington at 10:45 p.m. We stayed that night and started on in the morning getting into very hilly country immediately. The national pike, the Old Cumberland pike road, is certainly a wonderful road, but it does take one through very mountainous country. We crossed the Monongahela River at Brownsville and headed up into the mountains just east of Uniontown. Between Uniontown and Cumberland, which is 63 miles, we crossed 6 ranges of the Appalachians. We passed Braddock's grave and Washington's old Fort Necessity as well as the battlefield of the Great Meadows.

The highest of the mountains we went over was Big Savage Mountain, whose height is 2,000 feet. From this mountain, we had a steady drop down to Frostburg and then a steady downhill to Cumberland, 11 miles further, which is only about 300 feet high.

We went through the narrows of the Wills Creek gorge just north of Cumberland. We started on from Cumberland at about 5:00 p.m., with Hancock, Maryland, the next town east. This was 38 miles, and in between were 3 big ranges of the Appalachians, of which the last, Sideling Hill, was probably worst. . . . .

This was about 1,600 feet high, and I believe is quite a famous hill as a barrier to westward travel. We reached Hancock about 9:00 and camped about 2 miles east of Hancock along the National Pike.

In the morning we started on and went over one more range, which was quite low, Fairview Mountain, then dropped down into the Cumberland Valley and left the National Highway near Hagerstown, cutting up to Greencastle, Chambersburg, and Shippensburg. We stayed at the house of the Mclean's and had a very enjoyable time. We had Rachael in the shop most of the day, having new brake bands put in. Alice drove us in her Hudson in the afternoon to a pond about 5 miles from Shippensburg, and Nancy, Alice, Walter [Dew] and I had a swim for an hour or so. It was my first swim of the year, except in Spearfish Canyon, which was really too cold to be called a swim. The water of the pond was quite warm, and I enjoyed it very much. Alice and Walter are quite good divers. I swam 2/3 the length of the pond and back making about a quarter of a mile. . . . .

Sunday morning at about 9:00 we pulled out from Shippensburg, and went back through Chambersburg and on to Gettysburg, through a gap in the Blue Ridge Mountains. We passed right through the National Cemetery at Gettysburg. From there we went to York and crossed the Susquehanna on a long bridge about 12 miles east of York. We went on to Lancaster and then drove to Malvern. . . . . We then drove on to Philadelphia and on out to the Roger's place. We had supper there and stayed till 9:00. . . . . We started on from there about 9:00 p.m., went back to Philadelphia and crossed the Camden, then drove to Trenton via Mt. Holly and arrived in Princeton about 1:00 – on the day exactly 3 months after the day we pulled out from Princeton in June.

17.1   Expense Record, 1921 South Dakota Geological Expedition

FIELD EXPENSES:

1 Pint Oil at Bridgewater, S. D.	.15	1 Emergency brake at Rapid City	.40	Tube patching material	.50
Knuckle and spark connection	.25	6 Wheel bolts at Rapid City, July 18th	.30	Oil and gas at Scenic August 5th	1.00
Lodging for 3 at Bridgewater	2.50	Garage man's time at Rapid City	.75	Bump's Grocery Bill August 5th	3.28
Fan belt at Mitchell	.50	Food bill at Rapid City	4.65	Bump's Grocery Bill August 6th	1.95
Gum and tape for radiator at Mitchell	.15	One new front frame for car	1.75	Bump's Grocery Bill August 7th	1.60
3 pints oil at Kimball, S. D.	.40	Supper for H. R. W. at Rapid City	1.15	Bump's Grocery Bill August 9th	1.70
2 Cans radiator cement Chamberlin SD	1.50	Lunches July 19th	1.15	Bump's Grocery Bill August 10th	1.15
7 gallons gas at Chamberlin S. D.	1.75	6 Batteries for car	3.00	Bump's Grocery Bill August 12th	1.40
1 Quart oil at Chamberlin	.30	Bolts for frame	.35	Jost's Garage Bill August 12th	18.80
1 Can for water at Chamberlin	1.50	One lantern	.75	Bread August 8th and 11th	.30
Ferry crossing of Missouri River	1.20	Hose for radiator	.50	Suppers August 10th	1.15
Lunch for Dew at Chamberlin	.70,	Fixing kerosene stove	.10	Lunches August 7th	1.00
Lunch for H. R. W. at Chamberlin	.55	Screws for seat of car	.10	Lunches August 12th	1.15
Supper/ breakfast, July 4th & 5th, Vivian	1.90	7 Gallons of gasoline July 21st	1.75	Nails for boxes	.60
Supper/breakfast July 5th & 6th, Vivian	1.80	3 Quarts of oil July 23rd	.50	Driving shaft sleeve	1.25
Hotel bill at Vivian	3.00	Radiator leak repair	.75	Cotton August 10th	1.25
7 Gallons gas at Draper	1.75	Six gallons gas July 23rd	1.50	Oil and gas August 9th	1.00
1 Pint oil at Draper	.15	1 Quart of oil July 23rd	.35	Oil and gas August 10th	1.50
Lunch for Dew and H. R. W. at Murdo	1.20	1 Box of cotton July 23rd	1.25	Oil and gas August 11th	1.25
1 Quart oil at Pronhe	.15	1 Box of cotton July 27th	1.25	Mouse Trap	.05
One new radiator at Kadoka, S. D.	23.00	Six gallons gas July 27th	1.50	1 Quart of oil August 10th evening	.25
Five Gallons gas at Kadoka	.25	3 Pints of oil July 27th	.40	Bump's Grocery Bill August 13th	.65

2 Radiator hose connections, Kadoka	.40	Five gallons gas July 29th	1.25	Bump's Grocery Bill August 15th	.95
Supper and Lunch for Dew at Interior	1.00	1 Quart of oil July 29th	.25	Bump's Grocery Bill August 15th pm	1.45
Supper and lunch for H. R. W., Interior	1.00	3 Gallons of Gas July 30th	.75	Bump's Grocery Bill August 16th am	.80
Food supplies at Interior	.58	1 Quart of oil July 30th	.25	Bump's Grocery Bill August 16th pm	1.45
Food from Mr. Bump at Scenic, July 7	10.00	Mr. Bump's grocery bill July 30th	36.81	Tacks and bolt for car	.25
Basin, pail and can opener	1.30	Suppers for Dew and H. R. W. July 29th	1.00	Staples for boxes	.15
Lemons	.35	Lunches for Dew and H. R. W. July 30th	.85	Gas and oil August 16th	2.00
Wood alcohol	.25	Suppers for Dew and H. R. W. July 30th	1.00	Gas and oil August 15th	1.00
Bolts for wheel	.25	Grease July 30th	.50	Deposit on Goodyear tire August 16th	5.00
Suppers July 7th and 8th.	2.40	Breakfast for Dew & H. R. W. July 31st	1.00	Suppers August 15th	1.00
Ice Cream July 8th	.40	2 Gallons of gas at Rapid City	.50	Lunches August 16th	1.15
Suppers July 8, 10, 11 & lunch July 11	5.65	1 Quart of oil at Rapid City July 31st	.25	Oil and gas August 18th	2.08
Food bill from Mr. Bump	6.20	5 Gallons of gas at Custer	1.25	Jack for car August 18th	2.00
Syrup July 11	.55	3 Pints of oil at Custer	.50	Bread and eggs August 18th	.35
Repairing canteen	.75	Phone call at Custer	.10	Doughnuts August 19th	.30
Five gallons of gas July 9	1.25	Lunches at Custer August 1	1.10	Lunch August 19th at Interior	1.00
One pint of oil July 9	.15	Suppers at Lead City August 1st	1.05	Suppers. August 19th at Scenic	1.15
One inner tube at Murdo	2.75	Breakfasts at Lead August 2nd	1.00	Bump's Grocery Bill August 20th	6.45
Food bill July 14	11.70	Lunches at Lead August 2nd	.90	Paid to Earl Taylor for use of teams & wagon in July	20.00
Supper for Dew July 14	.90	Suppers at Deadwood	1.05	Milk from Taylors till August 23rd	3.10
Supper for H. R. W. July 14	.90	Breakfasts at Sturgis August 3rd	.55	Bump's Bill August 24th	6.45
Suppers for Dew and H. R. W. July 16	1.00	New inner tube at Lead	3.35	Cotton August 20th	1.20

2 Cans of milk July 16	.35	Photographic Plates at Lead	5.00	Cotton August 24th	1.25
Lunches for Dew and H. R. W. July 17	.50	Oil at Deadwood	.25	Shellac August 24th	.75
Suppers for Dew and H. R. W. July 17	1.00	Gas and oil at Rapid City	1.95	Pears and Bananas August 24th	.55
6 Gallons gas July 17	1.50	Lunch at Rapid City Aug 3rd	1.35	Gas and oil August 20th	2.25
1 Quart oil July 17	.25	Groceries at Rapid City	3.40	Gas and oil August 24th	1.75
Six gallons gas July 18 at Rapid City	1.50	New front spring & installation at Rapid City	5.50	Balance of money for Goodyear Tire	11.00
1 Quart oil July 18 at Rapid City	.25	Lens for light	.50	Supper August 23rd	1.15
1 Hub at Rapid City July 18	2.50	Oil and Gas at Miller, S. D.	1.62	Gas and Oil at Washington, Pa.	1.45
Bump's bill August 27th	2.25	Oil and Gas at Huron, S. D.	.75	Crank Handle Pin at Washington, Pa.	.15
Oil and gas August 29th	1.25	Suppers at Huron, S. D.	1.00	Oil at Uniontown, Pa., 2 Quarts	.50
Cotton August 29th	1.25	Breakfasts at Huron, S. D.	.95	Oil and Gas at Cumberland, Md.	1.75
Repairing of Water Bag	.10	Gas and Oil at Arlington, S. D.	.60	Suppers at Hancock, Md.	.85
Bump's Grocery Bill August 29th	.55	Lunches at Watertown, S. D.	1.35	Breakfast at Chambersburg, Pa.	1.00
Oil and gas August 30th	1.00	Gas at Sinceton, S. D.	1.25	Oil and Gas at Chambersburg, Pa.	2.25
Oil and gas at Keystone August 31st	.90	Oil at Effington	.15	Brake Bands Shippensburg, Pa.	3.00
Food Bill at Keystone	.50	Cylinder head gasket at Effington	.50	Oil at York	.50
Film Packs at Rapid City	1.80	Gasket and installation at Wahpeton	1.85	Oil and Gas at Lancaster, Pa.	2.00
Food bill at Hill City	.80	Oil and Gas at Fargo	2.25	Oil at Philadelphia, Pa.	.40
Suppers at Custer	1.00	Lunches at Fargo	1.00	Developer and hypo for developing summer pictures	1.10
Food bill at Custer	1.50	Oil at Fergue Falls	.25	**Subtotal**	**$ 529.83**
Shellac at Rapid City	2.70	Gas and Oil at Alexandria, Minn.	1.62	EXPENSES NOT INCURRED IN THE FIELD:	
Food at Rapid City	2.00	Oil and Gas at Minneapolis, Minn.	2.25	Purchase of 1916 Model Ford Car	$250.00
Oil and Gas at Rapid City	1.50	1 Quart of Oil at Sauk Center, Minn.	.25	Expedition's share of Public Liability	

Lunches at Rapid City	1.00	Suppers. at Alexandria, Minn.	.25	Insurance on car	21.00
Suppers at Scenic	1.10	Breakfast at St. Cloud, Minn.	.25	State License fee for car	7.50
Bump's Grocery Bill Sept. 1st	1.50	Breakfasts & suppers at Eahpeton	2.40	Repairs and permanent improvements to car during May and June	11.75
Bump's Grocery Bill Sept. 2nd	1.15	Lunches at Minneapolis, Minn.	.70	Tire Pump	3.00
Bump's Grocery Bill Sept. 3rd	1.65	Suppers at Menomonie, Wis.	1.00	Partial payment new Firestone tire	13.00
Putting in tie bolt and Tire Dough	1.85	Oil and Gas at Eau Claire, Wis.	2.25	Driver's License fee for H. R. Wanless	3.50
Suppers Sept. 3rd	1.10	Breakfasts at Eau Claire, Wis.	.70	Bill to Kingston Garage for putting car in shape for Western trip, including two new tires, new Wheel and 8 gallons gas, cleaning carbon, grinding valves and minor repairs	88.81
Bump's Grocery Bill Sept. 5th	2.35	Lunches at Shamrock, Wis.	.45	Summer Salary to H. R. W.	300.00
Cotton Sept. 5th	1.25	Oil at shamrock, Wis.	.25	**TOTAL OF EXPEDITION EXPENSES.**	
Oil and Gas Sept. 5th	1.25	Grease installed at Tonah, Wis.	.50	TOTAL FIELD EXPENSES	$ 529.83
Clip for Back Spring	.30	Gas at Tonah, Wis.	1.25	TOTAL PRELIMINARY EXPENSES AND SALARY	695.56
Shellac Sept. 5th	.75	Oil and Gas at Baraboo, Wis.	1.75	**TOTAL EXPENSES**	**$1,225.39**
Suppers Sept. 6th	1.00	Suppers at Madison, Wis.	.95	Less money paid in by Mrs. R. E. Wanless	−35.00
Bump's Grocery Bill Sept. 7th	1.20	Oil and Gas at Janesville, Wis.	1.50	**NET EXPENSES, 1921**	
Nails Sept. 7th	.55	Oil at Huntley, Ill.	.25	**GEOLOGICAL EXPEDITION**	**$1,190.39**
Nails Sept. 8th	.85	Gas in Elgin, Ill.	1.25	MONEY ADVANCED, 1921 EXPEDITION	
Wire Sept 8th	.50	Oil and Gas at Chicago, Ill.	1.64	Marsh Fund, Nat. Acad. Of Sci.	$400.00
Lunches Sept. 8th	1.00	Oil at La Porte, Indiana	.20	Princeton Geology Dept.	$500.00
Lunches Sept. 7th	.60	Blow Out patch at South Bend, Ind.	.50	By Professor W. J. Sinclair	$300.00
Mr. Jost's Bill for car Sept. 8th	7.00	Gas at Niles Mich.	1.50	**TOTAL ADVANCED**	**$1,200.00**
Welding Rod Sept. 8th	.85	Tire and Tube at Goshen, Ind.	31.75		
Lumber, staples, and paint brush	9.85	Oil at Goshen, Ind.	.30		
Oil and Gas Sept. 9th	1.50	Oil and Gas at Bryan, Ohio	1.04		
Milk bill for end of summer	1.00	Coffee at Bryan and Toledo, Ohio	.55		
Paint	.50	Gas and Oil at Toledo, Ohio	1.14		

Repairing of Water Bag	.10	Ferry for car to Sandusky	1.30	
Oil at Wall, S. D.	.15	Breakfasts at Sandusky	.80	
Oil at Phillip, S. D.	.25	3 Pints of Oil at Sandusky	.35	
Provisions at Phillip, S. D.	.55	Gas and Oil at Cleveland, Ohio	1.50	
Oil and Gas at Fort Pierre	1.80	Gas and Oil at Salem, Ohio	2.13	
Breakfasts. at Pierre	.70	Brake Band installed at Salem, Ohio	4.50	
Lunches for at Blunt	1.00	Oil at East Liverpool, Ohio	.35	
Connecting rod installed & oil at Blunt	4.60			

References

Sinclair WJ (1921a) A new *Hoplophoneus* from the *Titanotherium* beds. Proc Am Philos Soc 60:96–98

Sinclair WJ (1921b) The "turtle-Oreodon Layer" or "red layer," a contribution to the stratigraphy of the White River Oligocene, results of the Princeton University 1920 expedition to South Dakota. Proc Am Philos Soc 60:456–466

Sinclair WJ (1921c) Entelodonts from the Big Badlands of South Dakota in the Geological Museum of Princeton University. Proc Am Philos Soc 60:467–495

Wanless HR (1921a) Fossil collecting by motor in the Badlands. The Princeton Alumni Weekly 22(12):269–270

Wanless HR (1921b) Stratigraphic Notes, 1921 South Dakota Expedition, in Wanless, Harold R. 1921 Princeton Expedition to South Dakota, Notes and Photographs. Archived as VPD.00028 123050, Yale Peabody Museum of Natural History, New Haven, p 1–65

1922: Completing an Ancient Story (and a Few More Skulls)

Abstract

For the final summer of field research and collecting, Harold rode west with Professor Sinclair and Irish geologist, Thomas B. Lawler. Wanless was constantly addressing the meaning of geologic observations for reconstructing the ancient environments represented in the sediment sequences. Wanless completed mapping the fossil river channel deposits at the different levels in the Badlands' (White River) sequence. The group also evaluated the significance of limestone units with algal laminations within terrestrial rocks, the source for the volcanic ash deposited characterizing especially the upper part of the White River sequence, the origin of chalcedony veins cutting the strata, and causes for reddish to greenish to cream colors to the rocks. A summary of trip expenses is included.

Keywords

Wanless Butte · Rock colors · River channel deposits · Volcanic ash · Thomas Lawler

During the academic year of 1921–1922 Harold took graduate classes at Princeton, acted as a reader of student papers (getting paid for his efforts), and working on the descriptions of the rock samples that he had collected from the Badlands in 1920 and 1921. He passed his written preliminary exams in May and received funding for the following year. He wrote and had published a short note for the *American Mineralogist* in May 1922 about the sand crystals from Rattlesnake Butte (Fig. 17.9) collected in the summer of 1921 (Wanless 1922a). This was the first paper published by Wanless that discussed a portion of the geology of the Badlands region.

Harold Wanless returned to the Badlands in June of 1922 with Professor Sinclair and a visiting Irish geologist Thomas B. Lawler[1] (Fig. 18.1). They worked primarily in the Corral, Little Corral, Quinn, and Battle Creek draws area adjacent to Cooney Table (Figs. 17.4 and 18.2).

The goal of the 1922 expedition was to acquaint Sinclair with fossil localities from which he had not previously collected. The presence of Sinclair and Lawler making joint geologic observations followed by discussions and interpretations by all three proved to be invaluable. The results of these discussions were presented in Wanless (1923), his final and most comprehensive publication on the White River beds of the Badlands.

---

[1] Lawler would enter Princeton as a geology doctoral student in the Fall of 1922. He died in a boating accident in June 1923 while working in northern Quebec for the Canadian Geological Survey.

Fig. 18.1 (**a**) William
J. Sinclair, (**b**) Harold
R. Wanless, and (**c**)
Thomas B. Lawler in the
White River Badlands in
the summer of 1922.
(From Wanless (1922c),
archived by Yale
Peabody Museum)

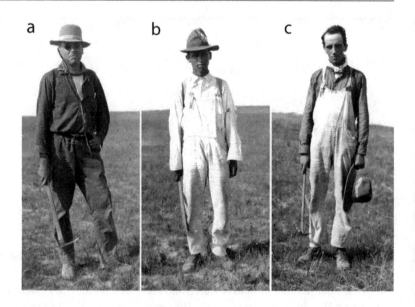

Fig. 18.2   Section of the edge of Cooney Table in Battle Creek Draw, showing heavy sandstones above Upper Nodular
Layer, upper *Oreodon* clays, and lower *Leptauchenia* beds (steep upper third of slope)

He expanded his research by describing and measuring the White River rocks on one of two large buttes on the western end of the Badlands Wall 5 miles northeast of the Imlay train stop (Figs. 17.4 and 18.3). This stratigraphic section provided information on the White River stratig-raphy within the 30 miles between the Scenic area in the west and the Interior area in the east.

Wanless completed mapping the patterns and characteristics of the river channels at the differ-ent levels in the sequence (Fig. 17.4). Thorough documentation of the river origin of portions of the sequence were important since the earliest

Fig. 18.3 Modern image looking north of the two large buttes on the west end of the Badlands Wall, 5 miles northeast of the Imlay train station. Wanless measured the sequence of rocks on the western flat-topped butte (on the left). These buttes are now informally named the Wanless Buttes in Badlands National Park. (Photograph by Emmett Evanoff)

interpretations had considered the sequence mostly lacustrine (lake-bed) deposits (Owen 1852; Hayden 1869). However, Matthew (1899) interpreted the White River rocks being derived mostly from eolian (wind) depositional processes, and Hatcher (1902) argued for a fluvial (river and stream) origin for these rocks. Wanless was recognizing all three – riverine (channel and floodplain), water laid (algal limestone ponds), and aeolian (wind transported dunes, desert-derived dust, and volcanic ash), and he was striving to document everything from the fine grain detail to the large-scale patterns of sand channels and aqueous, limestone-forming ponds (Wanless 1922b).

The three discussed what caused the changes in color (reddish to greenish to creamcolored), grain size of the rocks, the scale and lateral expanse of sandstone channel fills, and the composition and abundance of volcanic ash in the various units. All of this was done to try to differentiate the role of fluctuations of ancient climates, changes in the tectonic setting (uplift and basin down-dropping), and distant volcanic activity. Lawler and Wanless discussed the color banding in the *Titanotherium* mudstones as alternating wet (green beds) and dry (red beds) reflecting climatic cycles. They were developing not only an understanding of different stages of iron oxidation but also an appreciation that iron oxidation states that formed in the sediments a long ago can be preserved.

Wanless and his colleagues studied and discussed the various carbonate layers in the White River rocks. Some were massive discontinuous nodules aligned in layers indicating horizons of greater cementation (Fig. 1.7). The three researchers discussed how these could be caliches, or calcium carbonate deposits created when groundwater is drawn to the surface and evaporates leaving carbonate in the sediment. This was thought to have occurred not long after the sediment was deposited because of the presence of eroded and worn limestone nodules in the gravel of sandstone channel deposits and the lack of crushing by sediment compaction of occasional skulls that occur in the nodules.

Other limestone units were limestone beds containing freshwater invertebrate fossils (mostly freshwater snail shells), and some of these had algal laminations coating the shells. These were concluded to be freshwater pond deposits.

The White River rocks are cut by fractures filled either with sediment resulting in clastic dikes, or are filled by white or bluish gray chalcedony, a fibrous form of microcrystalline quartz. The clastic dikes can be filled with sand, silt, clay, or volcanic ash and then turned to stone.

Wanless and his field colleagues struggled with the origin and meaning of these fractures and their fillings. There were two main types of fracture fills – clastic sandstone/siltstone/claystone/volcanic ash or precipitated chalcedony. The clastic sediment fillings can be large and dramatic. They are well cemented and support long linear ridges and can extend for as much as a mile in length (Figs. 18.4 and 5.11). At the Cooney Table section "The dike filling in the upper *Leptauchenia* ash beds is a soft red clay. A soft red clay also filled the dike in the [overlying] *Protoceras* channel" (Wanless 1922c, note XXIII), suggesting fracturing and sediment infilling during more oxidized, arid conditions.

Whitish to bluish chalcedony (a microcrystalline quartz which includes onyx and agate) fills other fractures cut through a variety of lithologies (Fig. 18.5). Although these chalcedony-filled fractures are usually only an inch of so wide, they can be as much as 3 or 4 inches in the middle Oreodon beds near Cedar Butte (Wanless 1922c, note XXII). The chalcedony vein fills weather into thin vertical plates that separate and can litter the ground with plates of silica (Fig. 18.5).

Wanless and his colleagues contemplated a biogenic origin of the chalcedony – a dissolution and reprecipitation of unstable, opaline (non-

Fig. 18.4 Vertical clay ridge in middle of Oreodon Beds. [held up by a resistant sandstone dike], and another sandstone dike along the side of it, cutting it at an angle, near head of Battle Creek Draw. Sandstones at base are between upper and lower nodular layers. Person for scale. Black/white areas are two folded back negative emulsion tears. (From Wanless (1922c), archived by Yale Peabody Museum)

Fig. 18.5 Two views of silica vein-fills also known as chalcedony dikes. (**a**) A typical vein-fill with small but resistant dark gray silica wall. Plates of silica litter the slope on the right side of the image. Corral Draw. (**b**) Sandstone dike [white, about width of hat brim) lined on the right side with a [thin] chalcedony vein. Battle Creek Draw. (Photographs from 1922 expedition. Courtesy of the Yale Peabody Museum)

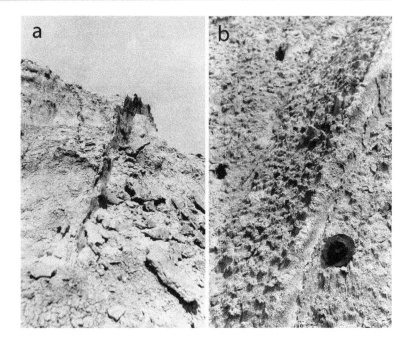

crystalline) silica possibly from skeletons of diatoms. They did not associate the also unstable and very abundant volcanic glass ash as a possible origin for the chalcedony in the fractures and the petrified wood (though he does in Wanless (1923)).

Plant fossils are rare in White River rocks, but Earl Taylor told the researchers about a petrified wood locality that he had found in the *Titanotherium* beds at the head of Corral Draw. The wood was part of a tree trunk and was partially carbonized and partially filled with blue chalcedony. The party visited and documented the site.

While Wanless and Lawler discussed the variations in the geologic features and their interpretations, Sinclair continued to add a substantial number of vertebrate fossils to the Princeton collection, including several titanothere skulls and the skull of a rare distant relative of the hippopotamus (an anthracothere).

18.1   Record of Expenses of 1922 Expedition to South Dakota Badland District

1.	E. H. Taylor for camp tending (Voucher 1)	$150.00
2.	Collecting sacks, rice paper, tools, shellac	14.97
3.	Motor supplies and repairs – other than gasoline and Oil – outbound trip (covered in part by Voucher 2)	59.69
4.	Lodging (outbound trip)	14.50
5.	Food (outbound trip)	14.34
6.	Fuel and lubricant (outbound trip)	33.90
7.	Sundries (outbound trip)	5.00
8.	Food consumed in camp (larger items covered by Vouchers.)	106.25
9.	Motor fuel in camp	8.90
10.	Sundries (with three items covered by Voucher 13)	21.81
11.	Food (return trip)	29.59
12.	Motor fuel and lubricants (return trip)	36.57
13.	Motor repairs – return trip (partly covered by Voucher 14)	11.75
14.	Lodging (return trip)	27.00
15.	Sundries (return trip)	5.59
	TOTAL	$539.87

References

Hatcher JH (1902) Origin of the Oligocene and Miocene deposits of the Great Plains. Proc Am Philos Soc 41:113–131

Hayden FV (1869) On the geology of the Tertiary formations of Dakota and Nebraska. In: Leidy J (ed) The extinct mammalian fauna of Dakota and Nebraska. Acad Nat Sci Philadelphia, 2nd Series 7, p 9–21

Matthew WD (1899) Is the White River Tertiary an eolian formation? Am Nat 33(389):403–408

Owen DD (1852) Incidental observations on the upper Missouri River, and descriptions of the geological formation of the Mauvaises Terres (Bad Lands) of Nebraska. Report of the Geol Survey of Wisconsin, Iowa, and Minnesota, Lippincott, Grandbo & Co, Philadelphia, Chapter 6, p 194–206

Wanless HR (1922a) Notes on sand calcite from South Dakota. Am Mineral 7(5):83–86

Wanless HR (1922b) Lithology of the White River sediments. Proc Am Philos Soc 61(3):184–203

Wanless HR (1922c) Field Stratigraphic Notes of 1922 South Dakota Expedition. Archived in Yale Peabody Museum of Natural History, VPD 00029 123051, 97p

Wanless HR (1923) The stratigraphy of the White River beds of South Dakota. Proc Am Philos Soc 62(4):190–269

Part IV

Wanless, Vertebrate Geology, and the Badlands Geology Since 1922

Part IV (Chaps. 19, 20, and 21)

Abstract The final three chapters look at both Harold Rollin Wanless and our understanding of White River Badlands geology in the century since 1920. Wanless became a faculty member at the University of Illinois and quickly applied his ideas of climate cyclicity from the Badlands to the late Paleozoic sedimentary rocks of the Midwest, recognizing ice-age, climate-driven cyclothems as an explanation for major repetitive sedimentary sequences. He emerged as a leading sedimentologist and deeply respected educator.

The physical difficulties in working in the White River Badlands in 1920 are still present and many of the techniques for collecting vertebrate fossils in 2020 are the same, but the materials, transportation, and amenities are different. The 1920s were the dawn of transportation by car, but much of the work in the Badlands had to be done by long hikes with limited water resources.

There have been significant advances in understanding the details of the stratigraphy, the fauna, and the ages of the Badlands sediments. Yet, as Sinclair stated, Wanless' work published in 1922 and 1923 was: "the best and most complete treatment of any Tertiary sedimentary problem that has ever been made in this country." It did set the standard for all subsequent work on White River and other North American Cenozoic rocks as Matthew stated. Wanless' two papers are still required reading for any modern geologist who wants to study the White River Badlands.

Abstract

Wanless became a faculty member at the University of Illinois in 1923. He applied his ideas of climate cyclicity that he had recognized in the Badlands rocks to the late Paleozoic sedimentary rocks of the Midwest. Wanless gained renown by recognizing ice-age climatic cycles as an explanation for major repetitive late Paleozoic sedimentary sequences that he named cyclothems. He emerged as a leading sedimentologist and deeply respected educator.

Keywords

Stratigraphy · White River sediments · University of Illinois · Grace Rogers Wanless · Cyclothems · Climate cycles · Aerial photography · Francis P. Shepard · Coastal changes

Wanless worked on and published his doctoral dissertation in two publications in the *Proceedings of the American Philosophical Society*. The first paper, *Lithology of the White River Sediments,* was published in 1922. The second, *The Stratigraphy of the White River Beds of South Dakota*, was published in 1923. Together these two publications are still considered to be the first modern analysis of the geology and paleoenvironments of the White River Group in the Badlands.

Wanless' detailed observations of the White River rocks and his integration of stratigraphy, sedimentary petrology, and paleontology to determine the paleo-environments of deposition and paleo-climate was far beyond what had been done previously. The stratigraphy paper received much praise after its publication. William Sinclair was very proud of his student's research and called his study "the best and most complete treatment of any Tertiary sedimentary problem that had ever been made in this country".[1] William D. Matthew, the curator of geology at the American Museum of Natural History and one of the preeminent authorities on Tertiary faunas in the United States in the early twentieth Century, stated that Wanless' paper "established a new standard for White River studies".[2] Herbert E. Gregory, chairman of the geology department of Yale University, wrote: "I am [especially] interested in your method and your recognition of the importance of the physical features of [the] sediments".[3]

Even to a modern sedimentary geologist the range of topics that Wanless discussed in his *Stratigraphy of the White River Beds* is impres-

---

[1] Letter to Rhoda Wanless, April 29, 1923.

[2] Letter to Rhoda Wanless, May 7, 1923.

[3] Letter to Harold Wanless, as recorded in a letter to Rhoda Wanless, November, 18, 1923.

sive. He discussed not only the subdivisions of the White River Group, but also the sedimentary petrology of the beds and used floral and faunal evidence for their origin. He discussed the origin of various nodules and concretions and the origin of clastic dikes,[4] evaluated the role of volcanic ash in forming silica cements; and speculated on the origin of the volcanic ash. He explored the post-White River geomorphic history of western South Dakota.

Even though the depositional models for river (fluvial), lake (lacustrine), and fine-grained wind deposits (loess) were just starting to be understood in 1920, Wanless did an admirable job in interpreting the origins of the various deposits in the Badlands. He also integrated the composition (lithology) of the rocks, their distribution in space and time, and their associated fossil plants and animals to interpret the paleoclimates during White River deposition. He discussed the possible effects that mountain building (tectonics) would have had on the deposition of White River sediments but concluded that paleoclimate variations were more important factors.

One of the most intriguing ideas that Wanless (1923, p. 249) briefly discussed was the concept that the deposition of the White River sediments showed repetitions (rhythmic changes) reflecting ancient paleoclimatic cycles. This was to become a feature of his later research in older Paleozoic-aged sediment sequences.

Upon completion of his Ph.D., he became a faculty member at the University of Illinois in 1923. He married Grace Rogers of Philadelphia in 1925 (Fig. 19.1). She was a lifelong support to his research and academic field training endeavors.

At Illinois, he was teaching the important new field of aerial photography interpretation to both geology and agronomy students. Through his career, he amassed a vast number of aerial photographs showing both geologic features, commonly in stereo, and also sequential changes in coastal environments.

As central Illinois was in the middle of vast coal deposits of late Paleozoic, Pennsylvanian Period, Wanless quickly became interested in the complex repeating sequences of sandstones, coal, limestone, and shales in which the coals were contained. He recognized that the repetitions of sediment types recorded a cyclic flooding and un-flooding of the continent as marine waters moved in and then withdrew yielding to coastal delta and terrestrial river environments. In 1932, Wanless and J. Marvin Weller published a description and interpretation of these Pennsylvanian sedimentary sequence marine to non-marine repetitions, calling them "cyclothems." They documented that individual cyclothems were widespread. Individual Pennsylvanian cyclothems were found to extend from Kentucky to Kansas indicating widespread ancient sea-level rises and falls, and these repeated rises and falls were on the order of 50 to 100 meters. They concluded that such dramatic and widespread repetitive changes in sea level and environments could only have been from global paleoclimatic controls. Specifically, repeated glacial buildup and retreat taking in water from and then releasing it back to the ocean was the only mechanism Wanless could imaging that could do this, and this was their interpretation.

It has stood the test of time. In 1932, we had only a vague idea of the dramatic sea level changes of the past million years, we had only spotty data of Pennsylvanian age glaciations, and we did not know about plate tectonics. In fact, with plate tectonics in the 1970s, we learned that a mega continent of Gondwana was sitting right on the South Pole in the Pennsylvanian and Permian. Made of Antarctica, Australia, India, South American and Africa, Gondwana records a coherent story of repeated polar ice advances and retreats. Until the proofs were clear, there were many challenges to what drove cyclothems. Tectonics turned out to be too regional and not able to generate many repetitions. Further proving the validity of the climate/glacial origin of cyclothems was a major component of Wanless' lifelong research, and it included meticulous study of outcrop and core-boring sequences, spatial mapping throughout the United States, and comparison with cyclothems throughout the

---

[4]Vertical cracks infilled by sediment, now understood to be the result of shrinkage as volcanic ash altered to clay minerals.

Fig. 19.1 Photograph of Grace Rogers Wanless and Harold Rollin Wanless taken at their wedding in 1925

world. In 1958–59 the family was in Australia with Wanless, a Fulbright Fellowship permitting him to focus on the Pennsylvanian and Permian cyclothems and glacial record.

The cyclothem concept driven by climate/ice cycles was Wanless' most significant contribution to sedimentary geology (Fig. 19.2), and its origin was in his early 1920s studies of the White River rocks in the Badlands.

His early focus on aerial photography also carried through his professional lifetime, primarily focusing on beach and coastal changes through time. Upon Wanless' retirement in 1967,

Harold and Grace spent much of a year in La Jolla, California producing a book, *Our Changing Shorelines*, with Francis Shepard of Scripps Institution of Oceanography. This 576 page effort documents and explains historical changes through sequential aerial photographs in the coasts throughout the United States (Shepard and Wanless 1971).

He was a valued and highly influential teacher and geoscientist at the University of Illinois throughout his life and one of the most influential sedimentary geologists of the mid Twentieth Century.

Fig. 19.2 Sign adjacent to the geology building at the University of Illinois commemorating the contribution Wanless made in defining and explaining cyclothems

References

Shepard FP, Wanless HR (1971) Our changing shorelines. McGraw-Hill Book Co, 576 p

Wanless HR (1922) Lithology of the White River sediments. Proc Am Philos Soc 61(3):184–203

Wanless HR (1923) The stratigraphy of the White River beds of South Dakota. Proc Am Philos Soc 62(4):190–269

Wanless HR, Weller JM (1932) Correlation and extent of Pennsylvanian cyclothems. Geol Soc Am Bull 43:1003–1016

Paleontological Field Work in 1920 and 2020

20

Abstract

The physical difficulties in working in the White River Badlands in 1920 are still present, and many of the techniques for collecting vertebrate fossils in 2020 are the same, but the materials, transportation, and amenities are different. The 1920s were the dawn of transportation by car, but much of the work in the Badlands had to be done by long hikes with limited water resources. Taxonomic names were still being resolved in the 1920s, but the illustrations of the skulls and reconstructions of the vertebrates by Horsfall in the early 1900s are still invaluable.

Keywords

Modes of travel · Heat · Badlands National Park · Brunton · Flour paste · Casting plaster · Shellac · Penetrating polymer consolidants · Global positioning systems · William Berryman Scott · R. Bruce Horsfall · Glenn L. Jepsen

Field work by bone diggers in 1920 had similarities to and major differences from field work a century later. One of the most profound changes was in in the mode of transportation.

Wanless travelled to the field by train, first taking the Chicago Northwestern Railway to Rapid City, South Dakota, and then the Chicago, Milwaukee, and St Paul Railroad to the towns of Scenic and Interior. At that time, the train was the fastest and most comfortable way to travel but it was much slower than today's modes of travel. It took an hour and a half to get from Rapid City to Scenic travelling at a rate of just over 30 miles per hour. By automobile, this trip today is a 45-minute drive on South Dakota Highway 44. Not all trains from Rapid City to Interior were passenger trains; Wanless had to catch a freight train from Scenic to Interior on June 27. That train did not stop in Interior but only slowed and Wanless had to jump off the slow-moving train and have his gear thrown to him, which was rather thrilling for the 21-year-old budding geologist.

The Sinclairs drove a Model-T Ford from Princeton, New Jersey, to Interior, South Dakota in 17 days, averaging about 108 miles per day. The roads in western South Dakota were all dirt in 1920, and in June of 1920, the roads were especially wet, making for very slow travel.

It took Sinclair four hours to drive the "George Washington Highway" from Interior to Scenic on June 9, 1920, a trip of 30 minutes on the modern South Dakota Highway 44. The Model T was an excellent field car because it was sturdy, had high clearance, and good maneuverability. In addition, it was built to withstand the roads of the time that were still mostly dirt wagon tracks. Although it was a relatively simple machine as compared to later automobiles, it took constant maintenance to keep the car running.

Cars were relatively new in 1920, but they revolutionized vertebrate paleontology. W. D. Matthew, a vertebrate paleontologist at the American Museum of Natural History, had experienced the transition between horse and car transportation. He wrote in 1926 (p. 454):

> The coming of the automobile has revolutionized the fossil-hunting business. For one thing, it has greatly widened the range of practical field work. The old conditions [horse transportation] limited it to a radius of five or ten miles from water and feed. A 'dry camp' supplied at intervals with food and water might carry the exploration a stage further, but through the West, there were enormous areas of bad-land exposures that [were] not practicable to prospect adequately for lack of water or feed. With the automobile, there is probably no promising exposure so distant but that it can be and will be prospected for fossils.

Sinclair had lived through this transportation transition and valued the advantages of the automobile. However, once he and Wanless were in the Badlands, cars provided only limited access to the outcrops, so in the field Sinclair and Wanless had to hike tens of miles each day to get to and from the bone beds.

Working in the White River Badlands has always been hazardous. The Badlands were named Makosica by the Lakota Sioux and Mauvaises Terres à Traverser by the French Canadians, both of which mean "bad lands to travel across." The badlands buttes, tables, and the Wall are all formidable barriers with extremely steep cliffs and knife-edge ridges with pinnacles. A hiker or climber in the Badlands can experience nasty falls on occasion. In some areas, a hiker has to walk miles around a set of badlands in order to cross the barrier where even game trails are few and far apart.

Wildlife in the Badlands can present challenges. Badlands National Park supports a bison population of over 1000 head, and a backcountry hiker must be vigilant to avoid these animals. Bison are not as docile as they sometimes appear and can easily outrun a human. Bison were not a problem for Wanless and Sinclair, because they had been hunted out of the region and were not

reintroduced into the area until the 1960s. Rattlesnakes are common but are easily avoided by the wary hiker and are seldom seen in the middle of a hot summer day.

The weather can be extreme in the Badlands, with prairie blizzards in the winter and local but very violent thunderstorms in the summer. Hail up to golf-ball (or larger) size can be a major problem for the exposed hiker or tent camper, and flashfloods are not uncommon. When wet, the White River claystone, mudstone, and siltstone beds become extremely slick and almost impossible to walk upon. However, the greatest danger to the hiker in the Badlands is the summer heat. The White River rocks and sediments are blinding white and create an intense reflector oven. Sunburn is a constant danger in these conditions, and sun hats, sunglasses, lightweight long-sleeve shirts, long pants, and ample sunscreen lotion is a must for the modern summer hiker. Temperatures in the Badlands can be as high as 120 °F (49 °C) and temperatures over 100 °F (38 °C) are common in the summer. In these extreme hot days, a hiker in the Badlands must drink between two gallons and four gallons of water per day. It is not a bad idea for modern researchers to seek shade and air-conditioning in the middle of the day and drink cool water and fluids with electrolytes. Modern researchers in the Badlands can stay in motel rooms or trailers with air conditioning to avoid developing heat exhaustion that builds from prolonged exposure.

None of these modern conveniences were available to Sinclair and Wanless in 1920. The Sinclairs stayed in a canvas wall tent and Wanless stayed under a canvas shelter-tent attached to the back of the car. Heavy canvas tents get very hot in the sun and retain the heat well into the night. Ventilation was better under the car tent, but it let in mosquitos and flies on warm nights.

Water was obtained from wells and springs in the area and brought to camp in a 10 gallon (37.8 L) metal can. While hiking in the field, Sinclair carried a three-pint (1.4 L) metal canteen and Wanless carried a quart (0.95 L) metal canteen. Both canteens heated the water to "a little short of boiling" on hot days according to Wanless' discussion on July 1. Wanless supplemented his water supply with an additional can of water later in the summer, but his total amount of available water in the field was still less than a gallon per day. After the very hot days of late July and early August, Delia Sinclair became ill and stayed more frequently in camp. William Sinclair also felt the effects of the heat, missing a day to sickness in early August and taking off from the field for the next two days. Although the cause of the sickness was not clear, the intense heat of the time suggests the Sinclairs were suffering from heat exhaustion.

By mid-August, Delia grew tired of the rigors of dryland camping and insisted that they return to Princeton two weeks earlier than they had planned. She may have also been concerned about the effect of the heat on her husband. Harold was less affected by the heat, perhaps because he was younger. He would go out in the field on day hikes alone while the Sinclairs stayed in camp. On one of these solitary excursions (on July 27) Wanless hiked a 30-mile (48 km) loop across the badlands, a remarkable feat given the small amount of water that he carried with him.[1] Wanless mentions drinking from water holes in Badlands streams, but a modern worker would not consider drinking such waters. The earliest researchers in the Badlands drank from water pools in the streams of the Badlands, and they and their horses would develop severe diarrhea (see the 1853 journal of Meek in Fryxell, 2010). Few springs in the area provided potable water.

The cost of field work was quite different in 1920 than 2020. In 1921 Wanless drove his own Model T to the Badlands and kept meticulous records of his expenses. The cost of gasoline in 1921 averaged 25 cents per gallon, and a quart of oil cost 25 cents per quart. The Model T used quite a bit of oil, 95 quarts or almost 24 gallons of oil for the season. Car repair and replacement parts were the biggest expense, totaling 32 percent of the total field expenses. Groceries cost an average of a little over $16.50 per week, and that fed Harold and two field assistants while camping in the field (averaging 79¢ per person per

day). Wanless and his field crew ate at restaurants frequently in 1921, and the price of breakfast averaged 44 cents, lunch averaged 52 cents, and dinner averaged 59 cents for a single meal. Lodging in hotels ranged from 88 cents to $1.50 per night per person in 1921.

Access to the lands has completely changed. In 1920, Every 40 acres of flatland in the Badlands was homesteaded and there were no public lands. Paleontologists and geologists were welcomed by the landowners to study the outcrops and collect fossils. No permits were required to collect fossils. In the late 1930s, the federal government bought up much of the homesteads as farms and ranches went bankrupt during the Dust Bowl. In 1939, Badlands was proclaimed a national monument by President Franklin Delano Roosevelt. The monument was expanded and made into a national park in 1978, including lands jointly administered with the Pine Ridge Indian Reservation. Much of the land surrounding the park is managed by the National Forest Service and was consolidated into Buffalo Gap National Grasslands in 1960. Research permits are now required for fossil collecting and geologic studies by professional paleontologists and geologists through the Park Service, National Forest Service, and the Pine Ridge Indian Reservation, depending on the area they are working.

Good relations with the local merchants, ranchers, and farmers were critical for the success of field work in 1920. Sinclair was particularly good at working with the ranchers and farmers, and Wanless continued the good relations for many years. The field crew relied on the local landowners for permission to camp, get water, and sometimes acquire such food as fresh milk and eggs. The most important residents of the area for the scientists during 1920 expedition included James Gay Bump and his family in Scenic; Walter Brown and his family near Interior; and Earl Taylor and David Motter on Hart Table. These people became friends with the Sinclairs and Wanless and provided them with a sense of community and companionship.

Wanless clearly described the field equipment and fossil excavation methods in his discussion of Fossil Hunting in the Badlands of South

---

[1] A modern hiker is not advised to replicate Wanless' hikes without carrying much more water.

Dakota (Chap. 1) and in the 1920 diary in the discussion of events of Tuesday, June 29. His discussion of field equipment reveals the field techniques used in 1920. The collecting sack, canteen, and pick were and are still standard items for collectors. Sinclair and Wanless carried newspaper, cotton, a shellac bottle, a shellac brush, a whisk broom, a curved digging tool (now an awl), a small iron chisel, and a small hammer. Again, these are current standard collecting materials and tools, although penetrating consolidants have replaced the shellac. The Brunton compass (Fig. 20.1) is a standard geologist's tool and is used for measuring the orientations of the long axis of bones, determining azimuths to landmarks, and the attitude (dip and strike) of inclined rocks. The compass can also be used to make detailed maps by measuring a baseline (by paces) and then triangulating between the baseline points and points of interest on the ground. Finally, the hand level was used to measure the thicknesses of the rocks, primarily by utilizing the eye-height of the researcher and using the hand level to sight the eye-height onto the rock sequence. Thick rock sequences were measured by summing multiple eye-height thicknesses while the thicknesses of thin beds were estimated, typically to the nearest foot (0.3 m).

The field techniques Wanless described are still followed by modern paleontologists (Fig. 20.2), but the materials have changed. Shellac is a resin secreted by the female lac bug (*Kerria lacca*) in the forests of Indonesia. It is a natural bioadhesive polymer that is soluble in alcohol and acts like a natural plastic that coats only the surface of the fossil. Unfortunately, shellac tends to strongly discolor fossils and degrades over time. Modern paleontologists use stable and penetrating polymer consolidants, such as Vinac™ (polyvinyl acetate) and Paraloid™ (an aqueous acrylic copolymer). Modern fossil jackets still use burlap or quilt-batting (light weight, strong, and easier to cut) soaked in casting plaster, covering a layer of paper or thin plastic sheeting that separates the plaster from the fossil. Flour paste and cloth were used in 1920 for fossil jackets because of their availability, but flour jackets were weaker, would mold over time, and were a favorite food of insects and mice.

In 1920, the only way to precisely document the exact location of a fossil site was by photography. Wanless did photograph some of the most important fossil localities, not only by photographing the fossils up-close, but by photographic the sites from a distance. These photographs have been invaluable in locating

Fig. 20.1 A vintage Brunton compass manufactured in 1914. It differs from a modern Brunton compass by having two level bubbles mounted at 90° from each other and no circular bulls-eye bubble. This is a model that Sinclair and Wanless may have used in the field in 1920

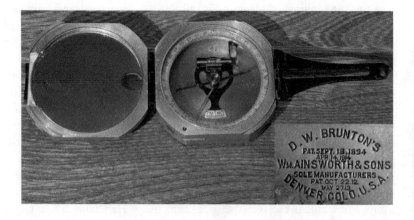

Fig. 20.2 Professor Sinclair in 1922 putting a flour-paste jacket on a rhinoceros jaw. The fossil is found, exposed, and pedestaled (rock removed from around the sides of the fossil) and then covered by the jacket. Between the jaw and the pick is a scrap of gunny sack that is mixed with the paste to help strengthen the jacket. Once the top jacket is dry, the base of the pedestal is cut, the fossil flipped over, and a basal jacket is added. The same procedure is used today only using plaster instead of flour paste. A second jacket can be seen drying behind Sinclair

such sites as the type locality for *Archaeotherium wanlessi*, Sinclair's specimen of *Archaeotherium clavus clavus*, and the oreodont skeleton at the Bear Creek Pocket. However, only a few fossil sites were photographed in detail. Wanless' sketch maps also provide information on where certain fossils were collected. However, detailed topographic maps of the Badlands in the form of 7.5-minute topographic quadrangle maps (scale 1:24,000) were not made for the Badlands until the 1960s, so exact plotting of localities on maps

were not possible in 1920. Modern paleontologists use global positioning systems (GPS) that use signals from satellites to precisely locate fossil sites to its latitude and longitude (in degrees, minutes, or seconds) or Universal Transverse Mercator positions (in meters). A handheld GPS device can locate a fossil site to the accuracy of ±10 feet (3 m), and advanced GPS devices can locate a fossil site to within a few centimeters. GPS was developed in the 1990s and became readily available in the 2000s, so such precision of locating fossil sites were not even dreamed of in 1920.

The taxonomic names of the White River vertebrate fossils had not been resolved in 1920. Fossil genera and species are named by taxonomic priority, or by the first name used in the first description of the fossil. For example, the fossil genus name *Merycoidodon* was first named and described by Joseph Leidy in 1848, but in 1851 he designated the name *Oreodon* to the same kind of fossil. The name *Merycoidodon* has priority and is the name current paleontologists use for this fossil genus. However, the genus *Merycoidodon* is in the Family Oreodontidae and are a member of the informal group called oredonts because the two genera cannot be determined to be exactly the same (McKenna and Bell 1997). Likewise, *Archaeotherium* is in the Family Entelodontidae, named for the European fossil genus *Entelodon*, and both pig-like animals are members of the informal group called entelodonts. In another instance, Sinclair used the name for the rhinoceros *Caenopus* named by Cope in 1880, which was later determined to be the same (synonymous) with *Subhyracodon* named by Brant in 1878. The name *Subhyracodon* has taxonomic priority.

The collection that Sinclair, Wanless, and other Princeton vertebrate paleontologists made of White River mammals was used by William Berryman Scott to make the five-volume monograph titled *The Mammalian Fauna of the White River Oligocene*, published between 1936 and 1941. This monograph helped to standardize the names of the White River fossil mammals. R. Bruce William Sinclair was to work on these monographs with William Berryman Scott, but Sinclair died in March of 1935. His successor, Glenn L. Jepsen worked with Scott as coauthors of the monographs.

R. Bruce Horsfall made detailed diagrams of all the White River mammal skulls for this monograph. Figure 20.3 shows these skull diagrams for most of the mammals collected by Sinclair and Wanless, published between 1936 and 1941. Sources of the skull illustrations are Scott and Jepsen (1936, Plates 8, 10, 13, 17, 19); Scott et al. (1940, Plate 34); Scott and Jepsen (1940, Plates 56, 60, 64, 69, 74, 76); Scott and Jepsen (1941, Plates 86, 89, 91, 96).

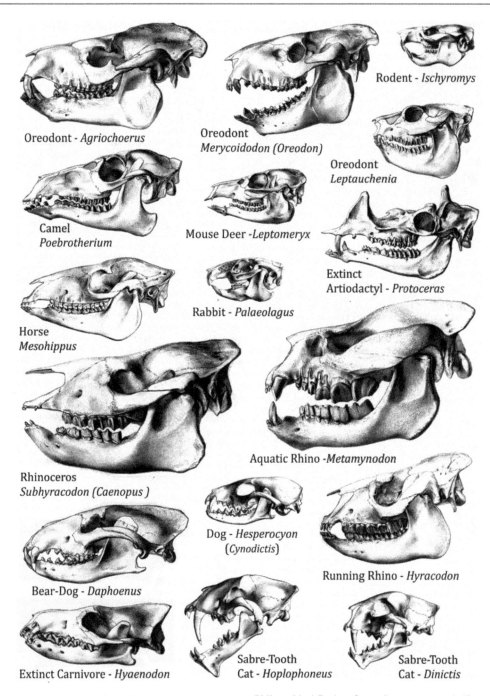

Fig. 20.3 Illustrations of skulls of most of the fossil mammals that were collected by Wanless and Sinclair in 1920. The names in parentheses are the names used by Wanless but are not used today. Illustrations are not to scale. The skull illustrations are from the American Philosophical Society five-volume monograph (Scott and others, 1936–1941, see text). (All illustrations were made by R. Bruce Horsfall. E. Evanoff assembled and labeled the Figure)

References

Fryxell FM (2010) Ferdinand Hayden, a young scientist in the Great West 1853–1855. Augustana Historical Society, Rock Island, 281p

Matthew WD (1926) Early days of fossil hunting on the High Plains. Nat Hist 26(5):449–454

McKenna MC, Bell SK (1997) Classification of mammals above the species level. Columbia University Press, New York, 631p

Scott WB, Jepsen GL (1936) The mammalian fauna of the White River Oligocene, part 1: Insectivora and Carnivora. Trans Am Philos Soc 28:1–152, plates 1–22

Scott WB, Jepsen GL (1940) The mammalian fauna of the White River Oligocene, part 4: Artiodactyla. Trans Am Philos Soc 28:363–733, plates 36–78

Scott WB, Jepsen GL (1941) The mammalian fauna of the White River Oligocene, part 5: Perissodactyla. Trans Am Philos Soc 28:734–980, plates 79–100

Scott WB, Jepsen GL, Wood AE (1940) The mammalian fauna of the White River Oligocene, part 3: Lagomorpha. Trans Am Philos Soc 28:271–362, plates 34–35

Abstract

There have been significant advances in understanding the details of the stratigraphy, the fauna, and the ages of the Badlands sediments (illustrated in Fig. 21.1). Yet, as Sinclair stated, Wanless' work published in 1922 and 1923 was: "the best and most complete treatment of any Tertiary sedimentary problem that has ever been made in this country." It did set the standard for all subsequent work on White River and other North American Cenozoic rocks as Matthew stated. Wanless' two papers are still required reading for any modern geologist who wants to study the White River Badlands.

Keywords

Stratigraphic framework · Correlation · Radioisotope dating · Red paleosol · Aeolian transport · River deposits · Volcanic ash deposits · Freshwater limestones · Riverine blanket mud deposits

The geologic framework (technically the stratigraphic framework or the understanding of the rocks and faunas in space and time) that Sinclair and Wanless were working with in 1920 is somewhat different than what we use today (Fig. 21.1). The rocks were named the White River "Series" by Meek and Hayden in 1858. These rocks were subsequently divided into three subdivisions by their vertebrate fossil content. Hayden (1858, 1869) named the lower *Titanotherium* beds and the overlying turtle-*Oreodon* beds. Wortman (1893) added an uppermost faunal subdivision, the *Leptauchenia* beds with the *Protoceras* channels and also recognized the *Metamynodon* channels in the turtle-*Oreodon* beds. Darton (1899) named two formations within the White River Group, the lower Chadron Formation and the overlying Brule Formation equivalent with the *Titanotherium* beds and the overlying turtle-*Oreodon* beds and *Leptauchenia-Protoceras* beds. Darton's type area for these two formations was in northwest Nebraska, but he extended the names to include the rocks in the White River Badlands of South Dakota. Early twentieth century vertebrate paleontologists still used the older faunal subdivisions of the *Titanotherium*, turtle-*Oreodon*, and *Leptauchenia-Protoceras* beds to subdivide the White River Group in the South Dakota Badlands. In 1920 the White River Group was considered to be Oligocene in age, from correlations between North American and European mammal faunas by Osborn and Matthew (1909).

The continental rocks of the western United States had been given formal names as formations by the start of the twentieth century. However, vertebrate paleontologists subdivided these rocks not by their changes in lithology, but by their distinct fossil mammal faunas. As a

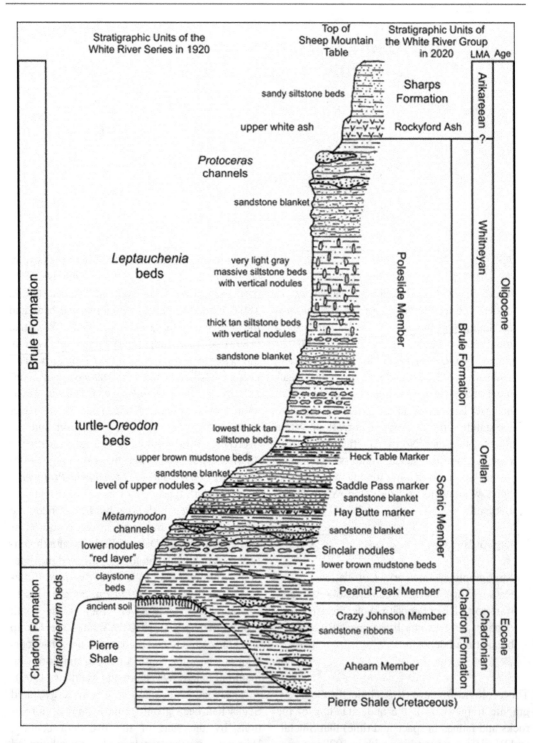

Fig. 21.1 The sequence of White River rocks at Sheep Mountain Table where Sinclair and Wanless worked in July and August of 1920. Shown are the names of the rocks (formations and members) and the faunal subdivisions of these rocks as recognized in 1920 and 2020. LMA are currently recognized Land Mammal Ages. The three marker beds in the Scenic, the Hay Butte, Saddle Pass, and Heck Table markers, are all widespread clayey mudstone units that allow for correlation across the entire region. The Hay Butte marker also contains a thin

result, the White River rocks of Badlands were divided into the *Titanotherium* beds, the turtle-*Oreodon* beds, and the *Leptauchenia* beds (Fig. 21.1). Similar mammal faunal subdivisions were established for the entire North American terrestrial rock record (see Osborn and Matthew, 1909, for an early list of these faunal subdivisions). These mammal faunal names were formalized as North American Land Mammal Ages by Wood et al. (1941) and these units represent biochrons, or a unit of time during which an association of taxa is interpreted to have lived (Woodburne 1977). Since fossil mammals typically occur in widely separated localities, and the formations in which they occur can be widely separated by geography, the order of land mammal ages (LMA) are determined by their stages of evolution (Tedford 1970).

The fauna of the *Titanotherium* beds were named by Wood et al. (1941) the Chadronian Land Mammal Age, the fauna of the turtle-*Oreodon* beds were named the Orellan Land Mammal Age, and the fauna of the *Leptauchenia-Protoceras* beds were named the Whitneyan Land Mammal Age. These names were derived from rock subdivisions of the White River Group in northwest Nebraska, the Chadron Formation, and the Orella and Whitney members of the Brule Formation. Unfortunately, the linkage of rock unit names with faunal unit names has been problematic because the lithologic boundaries of the rock members do not necessarily fall at the faunal changes. The upper-most rocks of the Badlands contain mammals from the younger Arikareean Land Mammal Age (Fig. 21.1), named for the Arikaree Group of Nebraska and South Dakota.

The geology of the Chadron Formation of the area west of Sheep Mountain Table was described by another Princeton graduate student, John Clark, in 1937. In 1954 Clark named three members of the Chadron Formation, the Ahearn, Crazy Johnson, and Peanut Peak members (Fig. 21.1). However, the three Chadron members can only be recognized in a paleovalley-fill west of Sheep Mountain Table. Outside the paleovalley fill the Chadron Formation is thinner and dominated by a thick sequence of claystone beds similar to the Peanut Peak Member (Fig. 21.1) but lateral to both the Peanut Peak and Crazy Johnson members (Clark et al. 1967).

In 1956, James Bump[1] named and described two members of the Brule Formation in the South Dakota Badlands. These were the Scenic and Poleslide members. In relation to the older faunal terminology, the Chadron Formation in the South Dakota Badlands is equivalent to the *Titanotherium* beds. The Scenic Member is within the lower and middle parts of the turtle-*Oreodon* beds. The Poleslide Member includes the upper part of the turtle-*Oreodon* beds and all the *Leptauchenia-Protoceras* beds. The upper white bed at Sheep Mountain Table, recognized by Wanless, is a lithified volcanic ash. This was named the Rockyford Ash by Nicknish and Macdonald in 1962. Wanless (1923) correlated this volcanic ash to a white bed in the Brule Formation in the Cedar Pass area by Interior, but subsequent work by Evanoff has shown these two white beds are separate units (Benton et al. 2015, Chap. 2). The Rockyford Ash is at the base of the Sharps Formation, named by Harksen, Macdonald, and Sevon (1961). The Sharps is composed of thick sandy siltstone beds that contain fossil mammals of the Arikareean Land Mammal Age, but the lowest occurrence of this fauna is uncertain on Sheep Mountain Table (Fig. 21.1).

_____

[1] Son of Gay Bump discussed in the 1920 diary.

◀

Fig. 21.1 (continued) volcanic ash bed. The "Sinclair nodules" is an informal name for the fossiliferous nodules at the top of the lower Scenic brown mudstone beds in the Entelodon Peak area west of Scenic. The total thickness of White River rocks from the base of the Chadron paleo-valley to the top of Sheep Mountain Table is 755 feet (230 m). From the Entelodon Peak area (on the left side of the rock column) the total thickness to the top of the mountain is 643 feet (196 m). The thicknesses of rock units in the Chadron paleo-valley are from Clark et al. (1967).

In 1920 Sinclair and Wanless collected fossils in two widely separated areas along the Badlands Wall. The Cedar Pass–Interior area is 35 miles east of Scenic and Sheep Mountain Table and they mis-correlated the rock units between these two areas. All the fossils they collected in the Interior area were in the upper Scenic and Poleslide members (middle turtle-*Oreodon* beds through the basal *Leptauchenia-Protoceras* beds) and none were low enough to be in the Chadron Formation (*Titanotherium* beds). In the Scenic and Sheep Mountain Table areas to the west, their most productive fossil beds were in the basal brown mudstones of the Scenic Member (lower turtle-*Oreodon* beds just above the *Titanotherium* beds) but they also collected from the Chadron Formation and the upper Poleslide Member.

The numeric ages of the White River rocks and faunas are a major difference between what was inferred in 1920 and what has now been determined. There were no radioisotopic dating techniques for sedimentary rocks in 1920. Today the faunas can be age-calibrated by dating volcanic ashes using either high-resolution uranium-lead analyses on the mineral zircon or by 40Argon/39Argon analyses on sanidine (a potassium-rich feldspar). As a result, the best age estimates for the White River rocks in the Badlands are now about 36 million years ago (Ma) for the base of the Chadron Formation to about 30 Ma for the top of the Whitneyan LMA (Benton et al. 2015). These estimates are much older than the estimate by Wanless of hundreds of thousands of years for the age of the rocks and fossils. These radiometric dates also indicate that the White River Group was deposited during the late Eocene through the early Oligocene.

A major difference between sedimentary geology in 1920 and today is the definitions of sedimentary rocks. Conglomerates and sandstones as understood then and now are essentially the same, but the terminology of mudrocks, or sedimentary rocks made of combinations of silt and clay (together called mud) are quite different. Part of the problem was that definitions of sedimentary particles was not standardized until 1922 by C. K. Wentworth. Wentworth's classification defined gravel as being particles larger than

2 mm; sand particles ranged from 1/16 mm to 2 mm; silt particles ranged from 1/256 mm to 1/16 mm; and clay was defined as particles less than 1/256 mm (4 microns) in diameter. Conglomerates are lithified gravel sediments and sandstones are lithified sand deposits. However, mudrocks are classified by texture and grainsize. All mudrocks that have very thin horizontal layers (called laminations) are shales. Many mudrocks that are blocky, or have no distinct internal laminations, and are classified by their grain size (Folk 1980). Thus, siltstones are blocky mudrocks with greater than 2/3 silt particles. Claystones are blocky and contain more than 2/3 clay particles. Mudstones are blocky and have intermediate clay/silt ratios, containing between 2/3 clay and 2/3 silt. All the mudrocks of the White River Group are claystones, mudstones, or siltstones.

The Chadron Formation is dominated by thick claystone beds surrounding scattered sandstone ribbons – narrow, long linear sandstone bodies. These sandstone ribbons were deposited by ancient river channels. The Scenic Member of the Brule Formation is dominated by two thick mudstone units over and under three widespread muddy sandstone blanket deposits. These sandstone blankets are composed of many stacked, thin sheets of sand deposited by streams flowing from the Black Hills and spread out over the plains. Two thin but widespread clayey mudstone beds separate the three sandstone blankets and are called the Hay Butte and Saddle Pass marker beds, regional time horizons. These two marker beds allow for correlation between the rocks at Sheep Mountain and the Cedar Pass area. Coarse sandstone ribbons scattered throughout the Scenic Member are equivalent to the *Metamynadon* channels of Wortman, Sinclair, and Wanless. In the west, the Scenic and Poleslide members are separated by a gray clayey mudstone marker bed called the Heck Table marker, but in the east the Scenic/Poleslide contact is marked by the change from mudstone beds into massive siltstone beds. The Poleslide Member of the Brule Formation is dominated by thick, massive siltstone beds, and secondary thick sandstone blanket deposits, composed of stacked thin sheets of silty sandstone. The Poleslide sandstone blankets contain rare coarse sandstone ribbons that are equivalent to the *Protoceras* channels.

Mudrock deposits of the White River Group were called by a variety of names in the early 1920s. Ward (1922) in his descriptions of the White River rocks near Interior called all the mudrocks "shales," though he noted that they did not contain laminations. Sinclair and Wanless refer to most of the mudrocks as "clays" for the Scenic mudstone beds and the lower siltstone beds of the Poleslide Member. The upper white siltstone beds of the Poleslide were called "ash beds" because of their high content of volcanic glass. Wanless called the middle Scenic sandstone blankets "siltstone beds" because he considered silt grains to be as large as 0.1 mm in diameter. Sediment particles 0.1 mm in diameter are now considered to be very fine sand in the Wentworth grain-size classifications. These differences in sedimentary rock nomenclature must be considered to compare the data gathered by Wanless to the observations of a modern sedimentary geologist.

Wanless (1923) was the first geologist to recognize the changes in depositional setting from the bottom to top (oldest to youngest rocks) within the White River Group. He recognized the erosional unconformity at the base of the White River Group developed on the underlying Cretaceous deposits, and the formation of a distinct red paleosol (an ancient soil and weathering horizon) under the contact. The basal *Titanotherium* beds were formed from the transition from an eroding landscape to one that became buried by deposition from streams. Valleys cut in the Cretaceous rocks were initially filled and then, as the old topography was buried, streams unconfined by valleys blanketed the land with widespread deposits. The greenish clays and local freshwater pond limestones in the *Titanotherium* beds indicated a seasonally wet-dry climate. A halt in sedimentation accompanied by some erosion occurred at the end of the deposition of the *Titanotherium* beds, and the erosional surface was then buried by sediment during the deposition of the turtle-*Oreo-*

don beds. Conditions were drier during turtle-*Oreodon* deposition, as indicated by brown to red sediments and the presence of abundant carbonate nodules that Wanless interpreted as soil carbonates (calcretes). Rivers still dominated the depositional systems, and many of these flowed out of the Black Hills onto a nearly featureless plain where the waters spread out and blanketed the surface with thin "siltstone" beds (now considered fine-grained sandstone beds). Finally, a large influx of volcanic ash blanketed the region during the deposition of the *Leptauchenia* beds. These volcanic ashes were deposited by wind and were interbedded with widespread blankets of thin-bedded silty sandstone and local ribbon sandstones of the *Protoceras* sandstones. The prevalence of aeolian (wind) deposits near the top indicated increased aridity from the sediment source areas at the end of White River deposition, as discussed by Wanless (1923).

The modern interpretation of the depositional history of White River rocks in the Badlands is similar to the interpretations of Wanless in 1923. We now know that the transition from erosion to aggradation at the start of Chadron deposition was caused by a great influx of volcanic ash that overwhelmed the streams and caused them to aggrade. The sources of the volcanic ash were much farther west than Wanless imagined, now known to be from huge and numerous late Eocene and early Oligocene volcanic eruptions in Nevada and Utah (Larson and Evanoff 1998). This ash during Chadron deposition was not only reworked by streams but also blanketed the topography and weathered into clays in a humid climate. After an interval of erosion where the Chadron deposits were cut into a low hill and valley topography, deposition resumed from addition of ash over the topography. Scenic Member streams flowing out of the Black Hills spread widespread blankets of stacked thin-bedded sand sheets over the subdued to flat topography of the plains. Conditions were drier as indicated by the preservation of volcanic glass shards in the mudstones of the Scenic Member and the presence of carbonate-rich paleosols.

Finally, the deposits changed from mudstone beds of the Scenic Member to thick siltstone beds of the Poleslide Member. These siltstone beds are widespread, contain little internal bedding, and contain abundant volcanic glass shards. The siltstone beds represent ancient dust deposits, called loessites. These deposits form when wind transports dust (silt-sized particles) from dry, arid desert settings to areas moist enough to support vegetation that can trap and hold the silt but dry enough so that there is little alteration of the volcanic glass shards into clay. The deposits of fluvial channels are rare and localized in these ancient fine-grained wind-derived deposits. Details of the depositional history of the White River rocks in the Badlands are given in Chap. 2 of Benton et al. (2015).

Wanless' work published in 1922 and 1923 was, as Sinclair stated, "the best and most complete treatment of any Tertiary sedimentary problem that has ever been made in this country.[2]" Itset the standard for all subsequent work on White River and other North American Cenozoic rocks as Matthew stated. Wanless' two papers are still required reading for any modern geologist who wants to study the White River Badlands.

References

Benton RC, Terry DO Jr, Evanoff E, McDonald HG (2015) The White River Badlands, geology and paleontology. University of Indiana Press, Bloomington, 224 p

Bump JD (1956) Geographic names for the members of the Brule Formation of the Big Badlands of South Dakota. Am J Sci 254:429–432

Clark J (1937) The stratigraphy and paleontology of the Chadron Formation in the Big Badlands of South Dakota. Carnegie Mus Ann 25:261–351

Clark J (1954) Geographic designation of the members of the Chadron Formation in South Dakota. Carnegie Mus Ann 33:197–198

Clark J, Beerbower JR, Kietzke KK (1967) Oligocene sedimentation, stratigraphy, paleoecology, and paleoclimatology in the Big Badlands of South Dakota. Fieldiana Geology Memoirs 5, 158 p

Darton NH (1899) Preliminary report on the geology and water resources of Nebraska west of the one hundred and third meridian. US Geol Surv Annu Rep 19(4):719–784

Folk RL (1980) Petrology of sedimentary rocks. Hemphill Publishing Company, Austin, 185 p

Harksen JC, Macdonald JR, Sevon WD (1961) New Miocene formation in South Dakota. Am Assoc Petrol Geol Bull 45:674–678

Hayden FV (1858) Notes on the geology of the Mauvais Terres of the White River, Nebraska. Proc Phila Acad Nat Sci 9:151–158

Hayden FV (1869) On the geology of the Tertiary formations of Dakota and Nebraska. In: Leidy J (ed) On the extinct mammalian fauna of Dakota and Nebraska. Philadelphia Acad Nat Sci J, 2nd Series (7), p 9–21

Larson EE, Evanoff E (1998) Tephrostratigraphy and source of the tuffs of the White River sequence. In: Terry DO Jr. LaGarry HE Hunt RM Jr. (eds) Depositional environments, lithostratigraphy, and biostratigraphy of the White River and Arikaree groups (late Eocene to early Miocene, North America). Geol Soc Am Spec Pap 325:1–14

Meek FB, Hayden FV (1858) Descriptions of new species and genera of fossils collected by Dr. F. V. Hayden in Nebraska Territory under direction of Lieut. G. K. Warren, U.S. topographical engineer, with some remarks on the Tertiary and Cretaceous formations of the north-west and the parallelism of the latter with those other portions of the United States and Territories. Proc Acad Nat Sci Phila 9:119–148

Nicknish JM, Macdonald JR (1962) Basal Miocene ash in White River Badlands, South Dakota. Am Assoc Petrol Geol Bull 46:685–690

Osborn HF, Matthew WD (1909) Cenozoic mammal horizons of western North America. US Geol Surv Bull 361, 138 p

Tedford RH (1970) Principles and practices of mammalian geochronology in North America. Proc North Am Paleontol Conv F:666–703

Wanless HR (1923) The stratigraphy of the White River beds of South Dakota. Proc Am Philos Soc 62(4):190–269

Ward F (1922) The geology of a portion of the Badlands. South Dakota Geol Nat Hist Surv Bull 11, Series 22(6), 80 p

Wentworth CK (1922) A scale of grade and class terms for clastic sediments. J Geol 30:377–392

Wood HE II, Chaney RW Jr, Clark J, Colbert EH, Jepsen GL, Reeside JB Jr, Stock C (1941) Nomenclature and correlation of the North American continental Tertiary. Geol Soc Am Bull 52:1–48

Woodburne MO (1977) Definition and characterization of mammalian chronostratigraphy. J Paleontol 51:220–234

Wortman JL (1893) On the divisions of the White River or lower Miocene of Dakota. Am Mus Nat Hist Bull 5:95–105

---

[2]From an unpublished letter from Wanless to his mother, April 29, 1923.

Appendix

List of fossils collected during the 1920 Princeton expedition. Courtesy of Mr. W. J. Sinclair
The original specimens can now be viewed at the Yale Peabody Museum (YPM) vertebrate paleontology
website at https://peabody.yale.edu/explore/collections/vertebrate-paleontology. Enter the YPM# in
Search the Collections to see the specimen. PU Loc is Princeton University locality in the 1920s.

Original Specimen #	Original Identification	Location Notes, Collector, Date	YPM # /Current Identification
12501	*Mesohippus* sp.	Associated maxillary. PU Loc 1014A2. Well up toward top of lower concretion layers, in zone of red concretions. W. J. Sinclair, June 29, 1920	VPPU.012501 *Mesohippus bairdi*
12502	*Mesohippus* sp.	Front of skull with good dentition. PU Loc 1014A2 in same level as specimen 12501. H. R. Wanless, June 29, 1920	VPPU.012502 *Mesohippus bairdi*
12503	*Mesohippus* sp.	Practically complete skull from PU Loc 1014A2. From butte removed but a short distance from the two preceding specimens. In Section 24, T. 3 S., R. 18 E., Black Hills Meridian. H. R. Wanless, June 29, 1920	No specimen in collection
12504	*Oreodon culbertsoni*	PU Loc 1014A2. Greenish part of lower concretionary zone rather poor skull. W. J. Sinclair, June 28, 1920	VPPU.012054 *Merycoidodon culbertsoni osborni*
12505	*Mesohippus* sp.	Front of skull with complete premolar-molar dentition. PU Loc 1012A2. From frosted clay and so badly rotten. Some associated lower jaw fragments. W. J. Sinclair, June 30, 1920	VPPU.012505 *Mesohippus bairdi*
12506	*Mesohippus* sp.	Associated maxilla with last molar not fully erupted. PU Loc 1012A2. W. J. Sinclair, June 30, 1920	VPPU.012506 *Mesohippus bairdi*
12507	*Mesohippus* sp.	Lower jaws from greenish portion of lower nodular layer. PU Loc 1014A2. W. J. Sinclair, July 1, 1920	VPPU.012507 *Mesohippus bairdi*
12508	*Mesohippus* sp.	Associated maxilla with last molar not fully erupted. PU Loc 1012A2. W. J. Sinclair, June 30, 1920	VPPU.012508 *Mesohippus bairdi*
12509	*Mesohippus* sp.	One ramus of lower jaw. From reddish rusty part of lower nodular layer; H. R. Wanless, July 1, 1920	No specimen in collection
12510	*Mesohippus* sp.	One ramus of lower jaw from surface of nodule layer. PU Loc 1014A2, W. J. Sinclair, July 3, 1920	VPPU.012510 *Mesohippus bairdi*
12511	*Mesohippus* sp.	One ramus of lower jaw from upper part of red banded clays below *Leptauchenia* clays. PU Loc 1014A5. West of middle pass. Saddle Pass is west of Cedar Pass and is not in condition to travel in the saddle. H. R. Wanless, July 5, 1920	VPPU 012511 *Mesohippus* sp.

H. R. Wanless, E. Evanoff, *The Diaries of a Bonedigger*, https://doi.org/10.1007/978-3-031-25118-4

Original Specimen #	Original Identification	Location Notes, Collector, Date	YPM # /Current Identification
12512	*Protoceras celer*	Male skull from *Protoceras* sandstone, a mile or so west of Cedar Pass at the top of the Wall, north of Interior, Jackson County, S. D. PU Loc 1014A7. W. J. Sinclair, July 6, 1920	VPPU.012512 *Protoceras celer*
12513	*Steneofiber nebrascensis*	Good skull probably from the lower nodular layer of the *Oreodon* beds, PU Loc 1014A2. Presented by Walter F. Brown, proprietor of Section 34, T. 3 S., R. 18 E., Interior, South Dakota	VPPU.012513 *Palaeocastor nebrascensis* No image available.
12514	*Mesohippus bairdi*	Fine skull with articulated lower jaws in place in a nodule of red zone of PU loc 1014A2. Between Cedar Pass and Saddle Pass, from low butte at the base of the wall, about a mile west of cedar pass road, almost due north of Brown's house in Section 34. H. R. Wanless, July 6, 1920	VPPU.012514 *Mesohippus bairdi*
12515	*Hoplophoneus mentalis*	Left ramus of lower jaw and supposedly associated skeleton fragments. Upper part of *Titanotherium* beds, east branch of Indian Creek, adjoining Hart Table. PU loc 1015A1, upper part. New species. W. J. Sinclair, July 13, 1920	VPPU.012515 *Hoplophoneus mentalis* Holotype
12516	*Echinatemys nebrascensis*	Complete carapace and plastron, toward head of east branch of Indian Creek, south of Hart Table. PU loc 1015A2A. H. R. Wanless, July 13, 1920	VPPU.012516 *Stylemys nebrascensis*
12517	*Mesohippus* sp.	Associated lower molars, right and left. Upper part of the *Titanotherium* beds, east brach of Indian Creek adjoining Hart Table. PU loc 1015A1 (upper part). W. J. Sinclair, July 13, 1920	VPPU.012517 *Mesohippus* sp.
12518	*Hyracodon nebrascensis*	Good upper dentition in rather poor skull. Clay within nodular layer of lower *Oreodon* beds, toward the head of the east branch of Indian Creek south of Hart Table. PU loc 1015A2a. H. R. Wanless, July 13, 1920	VPPU.012518 *Hyracodon nebrascensis*
12519	*Ischyromys typus*	Jaws found together in clays above the nodular layer not far below the top of Hart Table. *Oreodon* beds, east branch of Indian Creek. PU loc 1015A2b. H. R. Wanless, July 13, 1920	VPPU.012519 *Ischyromys typus*
12520	*Leptomeryx* sp.	Lower jaws supposedly associated. Upper part of *Titanotherium* beds, east branch of Indian Creek, adjoining Hart Table. PU loc 1015A1, upper part. W. J. Sinclair, July 13, 1920	VPPU.012520 *Leptomeryx* sp.
12521	*Hyaenodon cruentus*	Splendid skull from 1015A2a, in a loose nodule undobtedly from the red turtle-*Oreodon* layer. Head of east branch of Indian Creek. W. J. Sinclair, July 14, 1920	VPPU.012521 *Hyaenodon (Neohyaenodon) horridus*
12522	*Archaeotherium wanlessi*	Splended skull and lower jaws with several neck vertebrae in a nodule from the red turtle-*Oreodon* layer, at the head of the east branch of Indian Creek. Southeast of *Entelodon* Peak. PU loc 1015A2a. H. R. Wanless, July 14, 1920	VPPU.012522 *Archaeotherium wanlessi* Holotype
12523	*Stylemys nebrascensis*	Fine carapace and plastron of medium-sized turtle from the turtle-*Oreodon* layer, PU loc 1015A2a, east branch of Indian Creek *Entelodon* Peak area. H. R. Wanless, 14 July, 1920	VPPU.012523 *Stylemys nebrascensis*

Original Specimen #	Original Identification	Location Notes, Collector, Date	YPM # /Current Identification
12524	*Oreodon* (large sp.)	Skull and lower jaws in nodule from the red turtle-*Oreodon* layer, head waters of east branch of Indian Creek. PU loc 1015A2a, lower *Oreodon* beds. H. R. Wanless, 15 July 1920	VPPU.012524 *Merycoidodon culbertsoni*
12525	*Oreodon gracilis*	Skull and jaws in a nodule from the red turtle-*Oreodon* layer, head of the east branch of Indian Creek, *Entelodon* Peak area. PU loc 1015A2a, lower Oreodon beds. W. J. Sinclair, July 15, 1920	VPPU012525 [*Miniochoerus gracilis*]
12526	*Oreodon gracilis*	Skull and jaws found with specimen 12525. From the red turtle-*Oreodon* layer, head of the east branch of Indian Creek, *Entelodon* Peak area. PU loc 1015A2a, lower *Oreodon* beds. W. J. Sinclair, July 15, 1920.	VPPU012526 [*Miniochoerus gracilis*]
12527	*Ischyromys* sp.	Doubtfully associated and dissociated maxilla and jaw fragments. From the red turtle-*Oreodon* layer, head of the east branch of Indian Creek. PU loc 1015A2a, lower *Oreodon* beds. H. R. Wanless, July 15 & 16, 1920. Found with *Palaeolagus* jaws, specimens 12601 and 12602	VPPU.012527 *Ischyromys typus*
12528	Egg	Replaced by chalcedony? Same locality as specimen 12527. From the red turtle-*Oreodon* layer, head of the east branch of Indian Creek, *Entelodon* Peak area. PU loc 1015A2a, lower *Oreodon* beds. H. R. Wanless, July 14, 1920	VPPU.012528 *Neornithes*?
12529	*Archaeotherium clavus clavus*	Specimen from shales of rusty nodular layer. West side of *Entelodon* Peak. Badlands near headwaters of most easterly branch of Indian Creek. PU loc 1015A2a, lower *Oreodon* beds. H. R. Wanless, July 16, 1920	VPPU.012529 *Archaeotherium mortoni*
12530	*Archaeotherium clavus clavus*	Large concretion specimen from 100 yards from the previous specimen (12529). From the red turtle-*Oreodon* layer, head of the east branch of Indian Creek. PU loc 1015A2a, lower *Oreodon* beds. H. R. Wanless, July 16, 1920	VPPU.012530 *Archaeotherium mortoni*
12531	*Colodon* sp.	Single upper molar from zone of rusty red concretions near base of *Oreodon* beds. 105A2a. West side of *Entelodon* Peak. W. J. Sinclair, July 16, 1920	VPPU.012531 *Subhyracodon* sp. No image available
12532	*Entelodon mortoni*	From the red turtle-*Oreodon* layer, head of the east branch of Indian Creek. PU loc 1015A2a, lower Oreodon beds, *Entelodon* Peak area. H. R. Wanless, July 15, 1920.	VPPU.012532 *Archaeotherium mortoni*
12533	*Mesohippus* sp.	Rather poor skull, badly broken, but with associated upper and lower teeth. From the red turtle-*Oreodon* layer, head of the east branch of Indian Creek, *Entelodon* Peak area. PU loc 1015A2a, lower *Oreodon* beds. H. R. Wanless, July 17, 1920	VPPU.012533 *Mesohippus* sp.
12534	*Stylemys nebrascensis*	Large and almost perfect shell. From the red turtle-*Oreodon* layer, head of the east branch of Indian Creek, *Entelodon* Peak area. PU loc 1015A2a, lower *Oreodon* beds. H. R. Wanless, July 13, 1920	VPPU.012534 *Stylemys nebrascensis*
12535	*Oreodon* sp.	Skull and lower jaws from red concretion layer, lower *Oreodon* beds, *Entelodon* Peak area. PU loc 1015A2a. W. J. Sinclair, July 17, 1920	VPPU.012535 *Merycoidodon culbertsoni*

Original Specimen #	Original Identification	Location Notes, Collector, Date	YPM # /Current Identification
12536	*Dinictis felina*	Skull and package of bone fragments from rusty concretionary layer near the base of *Oreodon* beds west side of Spring Creek basin adjoining Hart Mountain. PU loc 1015B2a. W. J. Sinclair, July 19, 1920	VPPU.012536 *Dinictis felina*
12537	*Oreodon* sp.	Skull and lower jaws from concretionary layer, Entelodon Peak area, PU loc 1015A2a. H. R. Wanless, July 19, 1920	VPPU.012537 *Merycoidodon culbertsoni*
12538	*Agriochoerus* sp.	Good skull from red concretionary layer. West side of Spring Creek basin adjoining Hart Mountain. PU loc 1015B2a. W. J. Sinclair, July 19, 1920	VPPU.012538 *Agriochoerus antiquus*
12539	*Hyaenodon cruentus*	Good skull and jaws with extreme front missing. Indian Creek drainage basin, west side of Hart Mountain, rusty concretion zone. PU loc 1015A2a. H. R. Wanless, July 20, 1920	VPPU.012539 *Hyaenodon (Neohyaenodon) horridus*
12540	*Hyaenodon crucians*	Skull of small species from a few yards from specimen 12539. Indian Creek drainage basin, west side of Hart Mountain, rusty concretion zone. PU loc 1015A2a. H. R. Wanless, July 20, 1920	VPPU.012540 *Hyaenodon crucians*
12541	*Mesohippus* sp.	Skull and a few foot and limb bones. Spring Creek pocket of nodular layer. W. J. Sinclair, July 20, 1920	VPPU.012541 *Mesohippus bairdi*
12542	*Dinictis* sp.	Skull and associated skeletal fragments. Clays immediately above brown nodule layer, south fork of east Indian Creek. PU loc 1015A2a. W. J. Sinclair, July 21, 1920	VPPU.012542 *Hoplophoneus primaevus*
12543	*Oreodon* sp.	Skull and lower jaws from red concretionary layer, headwaters of the most easterly branch of Indian Creek, lower *Oreodon* beds. PU loc 1015A2a. H. R. Wanless, July 23, 1920	No specimen in collection
12544	*Entelodon mortoni*	Young skull. Headwaters of most easterly branch of Indian Creek, lower *Oreodon* beds, PU loc 1015A2a. W. J. Sinclair, July 23, 1920	VPPU.012544 *Archaeotherium mortoni*
12545	*Oreodon* sp.	Skull. Headwaters of most easterly branch of Indian Creek, lower *Oreodon* beds, PU loc 1015A2a. W. J. Sinclair, July 23, 1920	No specimen in collection
12546	*Entelodon* sp.	Lower jaw of large species. From headwaters of most easterly branch of Indian Creek, from the yellow-weathering shales below the brown concretionary zone. W. J. Sinclair, July 23, 1920	VPPU.012546 *Archaeotherium wanlessi*
12547	*Ischyromys* sp.	Lower jaw from rusty concretionary zone. Headwaters of most easterly branch of Indian Creek, lower *Oreodon* beds, PU loc 1015A2a. W. J. Sinclair, July 23, 1920	No specimen in collection
12548	*Ischyromys* sp.	Poor skull from clays just above the rusty nodular zone of 1015A2a. Headwaters of most easterly branch of Indian Creek, lower *Oreodon* beds. H. R. Wanless, July 23, 1920	VPPU.012548 *Ischyromys typus* No image
12549	*Mesohippus* sp.	Poor skull from brown nodular layer of lower *Oreodon* beds in western part of Indian Creek basin in vicinity of Cedarcovered butte. PU loc 1015C2a. H. R. Wanless, July 25, 1920	VPPU.012549 *Mesohippus bairdi*
12550	*Entelodon* sp.	Lower jaws of large species from brown nodular layer on one of the projecting spurs of *Entelodon* Peak. PU loc 1015A2a. W. J. Sinclair, July 27, 1920	VPPU.012546 *Archaeotherium wanlessi*

Original Specimen #	Original Identification	Location Notes, Collector, Date	YPM # /Current Identification
12551	*Dinictis* sp.	Poor skull and associated skeletal fragments from nodular layer, east Indian Creek pocket. PU loc 1015A2a. W. J. Sinclair, July 27, 1920	VPPU.012551 *Dinictis felina*
12552	*Oreodon* sp.	Skull and jaws from brown nodular layer from western part of Indian Creek basin in vicinity of Cedar-covered butte. PU loc 1015C2a. W. J. Sinclair and H. R. Wanless, July 27, 1920	No specimen in collection
12553	*Hyaenodon crucians?*	Skull, jaws and skeletal fragments. Head very fine. Bear Creek pocket, PU loc 1015D2a. W. J. Sinclair, July 31, 1920	VPPU.012553 *Hyaenodon crucians*
12554	Turtle	Good turtle in two packages. Bear Creek pocket, PU loc 1015D2a. W. J. Sinclair, July 31, 1920	VPPU.012554 *Stylemys nebrascensis*
12555	Turtle	Good turtle in two packages. Bear Creek pocket, PU loc 1015D2a. W. J. Sinclair, July 31, 1920	No specimen in collection
12556	Turtle	Good turtle in two packages. Bear Creek pocket, PU loc 1015D2a. H. R. Wanless, July 31, 1920	No specimen in collection
12557	*Oreodon gracilis*	Skull and jaws. Bear Creek pocket, PU loc 1015D2a. H. R. Wanless, July 31, 1920	VPPU.012557 *Miniochoerus gracilis*
12558	*Dinictis felina*	Front of skull and half of lower jaw. Bear Creek pocket, PU loc 1015D2a. H. R. Wanless, July 31, 1920	VPPU012558 *Dinictis felina*
12559	*Oreodon* sp.	Skull and jaws in nodule. Bear Creek pocket, PU loc 1015D2a. H. R. Wanless, July 31, 1920	No specimen in collection
12560	*Oreodon* sp.	Skull and jaws in nodule. Bear Creek pocket, PU loc 1015D2a. W. J. Sinclair, July 31, 1920	No specimen in collection
12561	*Oreodon* sp.	Skull and jaws in nodule. Bear Creek pocket, PU loc 1015D2a. W. J. Sinclair, July 31, 1920	VPPU.012561 *Merycoidodon culbertsoni*
12562	*Oreodon* sp.	Skull and jaws in nodule. Bear Creek pocket, PU loc 1015D2a. Mrs. W. J. Sinclair, July 31, 1920	VPPU.012562 *Merycoidodon culbertsoni*
12563	*Hyracodon* sp.	Upper and lower teeth. Bear Creek pocket, PU loc 1015D2a. W. J. Sinclair, July 31, 1920	VPPU.012563 *Hyracodon nebrascensis*
12564	*Oreodon* sp.	Good skull and jaws in 3 packages. Bear Creek pocket, PU loc 1015D2a. H. R. Wanless, July 31, 1920	No specimen in collection
12565	*Oreodon* sp.	Skull and skeleton in nodule. Nine packages. Bear Creek pocket, PU loc 1015D2a. H. R. Wanless, August 2, 1920	VPPU.012566 *Merycoidodon culbertsoni*
12566	*Oreodon* sp.	Skull and jaws in nodule. Bear Creek pocket, PU loc 1015D2a. H. R. Wanless, August 2, 1920	VPPU.012565 *Merycoidodon culbertsoni*
12567	*Hyracodon* sp.	Lower milk dentition. Bear Creek pocket, PU loc 1015D2a. W. J. Sinclair, August 2, 1920	VPPU.012563 *Hyracodon nebrascensis*
12568	*Entelodon mortoni*	Skull and fragments. Bear Creek pocket, PU loc 1015D2a. H. R. Wanless, August 2, 1920	VPPU.02568 *Archaeotherium mortoni*
12569	Large turtle	Two packages. Bear Creek pocket, PU loc 1015D2a. H. R. Wanless, August 2, 1920	VPPU.012569 *Stylemys nebrascensis*
12570	*Oreodon* sp.	Skull and jaws in nodule with package of fragments. Bear Creek pocket, PU loc 1015D2a. H. R. Wanless, August 2, 1920	VPPU.012565 *Merycoidodon* sp.

Original Specimen #	Original Identification	Location Notes, Collector, Date	YPM # /Current Identification
12571	Small turtle	Complete. Bear Creek pocket, PU loc 1015D2a. W. J. Sinclair, August 2, 1920	VPPU.012571 *Stylemys nebrascensis*
12572	*Mesohippus* sp.	Front of skull and some fragments. Bear Creek pocket, PU loc 1015D2a. H. R. Wanless, August 3, 1920	VPPU.012572 *Mesohippus bairdi*
12573	*Dinictis* sp.	Lower jaws and scrap. Bear Creek pocket, PU loc 1015D2a. W. J. Sinclair, August 3, 1920	VPPU.012573 *Dinictis squalidens*
12574	*Oreodon* sp.	Skull and jaws. Bear Creek pocket, PU loc 1015D2a. H. R. Wanless, August 3, 1920	VPPU.012574 *Merycoidodon culbertsoni*
12575	*Oreodon* sp.	Skull and jaws. Bear Creek pocket, PU loc 1015D2a. W. J. Sinclair, August 3, 1920	VPPU.012574 *Merycoidodon culbertsoni*
12576	*Oreodon gracilis*	Skull. Bear Creek pocket, PU loc 1015D2a. H. R. Wanless, August 3, 1920	VPPU.012576 *Miniochoerus gracilis*
12577	*Dinictis* sp.	Lower jaw and parts of skeleton. Bear Creek pocket, PU loc 1015D2a. H. R. Wanless, August 3, 1920	VPPU.012577 *Dinictis felina*
12578	*Oreodon* sp.	Skull, jaws, and part of skeleton in a large nodule. Bear Creek pocket, PU loc 1015D2a. W. J. Sinclair, August 3, 1920	VPPU.012578 *Merycoidodon* sp.
12579	Peccary skull	From white ash layer of *Leptauchenia* clays, Sheep Mountain at Stony Pass between the main table and the western pinnacles. 2 packages. H. R. Wanless, August 4, 1920	VPPU.012579 *Perchoerus probus*
12580	*Hyaenodon cruentus*	Skull and a few skeletal fragments. Clay specimen just below the nodules (about 2 feet below). On Indian Creek side of Indian Creek-Corral Draw divide. PU loc 1015C2a	VPPU.012580 *Hyaenodon (Neohyaenodon) horridus*
12581	*Oreodon* sp.	Skull and jaws, fine white bone. Bear Creek pocket, PU loc 1015D2a, just below the upper band of nodules. H. R. Wanless, August 10, 1920	VPPU.012581 *Merycoidodon culbertsoni*
12582	*Oreodon* sp.	Skull and jaws, fine white bone. In Spring Creek drainage but inseperatable from the Bear Creek pocket as a continuous horizon. H. R. Wanless, August 10, 1920.	VPPU.012582 *Merycoidodon culbertsoni*
12583	*Oreodon* sp.	Skull and jaws. On Spring Creek side of buttes that separate Spring and Bear creek drainages. Same horizon as in the Bear Creek pocket, PU loc 1015D2a. H. R. Wanless, August 10, 1920	VPPU.012583 *Merycoidodon culbertsoni*
12584	*Poebrotherium* sp.	Limb bones and good foot. Bear Creek pocket, PU loc 1015D2a. W. J. Sinclair, August 10, 1920	VPPU.012584 *Poebrotherium* sp.
12585	*Poebrotherium* sp.	Poor skull and some skeletal parts. Bear Creek pocket, PU loc 1015D2a. W. J. Sinclair, August 10, 1920	VPPU.012585 *Poebrotherium* sp.
12586	*Mesohippus* sp.	Upper teeth and lower jaws. Bear Creek pocket, PU loc 1015D2a. Upper part of nodular layer. H. R. Wanless, August 10, 1920	VPPU.012586 *Mesohippus bairdi*
12587	*Daphoenus* sp.	Skull and fragments, dentition complete from lower p3 to m3. Spring Creek pocket, PU loc 1015B2a. Clays below the nodular layer. H. R. Wanless, August 11, 1920	VPPU.012587 *Daphoenus vetus*
12588	*Daphoenus* sp.	Toothless skull with fine brain cast from reddish clays of lower nodular layers. 4.5 miles east of Scenic on and to Interior at west headwaters of Jones Creek. Jones Creek is given on the county map as a branch of Cain Creek. PU loc 1015E2a. Mrs. W. J. Sinclair, August 12, 1920	VPPU.012588 *Daphoenus vetus*

Original Specimen #	Original Identification	Location Notes, Collector, Date	YPM # /Current Identification
12589	*Mesohippus* sp.	Skull and lower jaws from reddish clays of red concretionary layer, 13.4 feet above white silidified lens of limestone marking top of *Titanotherium* beds. About 200 yards to NE of stake marking section corner in Section 30, T. 3 S., R. 14 E. W. J. Sinclair, August 13, 1920.	VPPU.012589 *Mesohippus bairdi*
12590	*Hoplophoneus insolens*	Large skull and a good deal of skeleton and skeletal fragments. Headwaters of most westerly branch of Jones Creek, 4.5 miles east of Scenic From the brown nodular layer in the caliche nodules. PU loc 1015E2a. H. R. Wanless, August 13, 1920	VPPU.012590 *Hoplophoneus primaevus latidens*
12591	*Hyracodon* sp.	Forefoot complete, found with all the elements in articulation. West side of buttes south of 71 Table, supposedly Culbertson's locality. PU loc 1015F2a. W. J. Sinclair, August 14, 1920.	VPPU.012591 *Hyracodon* sp.
12592	*Caenopus* sp.	Hind foot, tibia. Femur not collected. West side of buttes south of 71 Table, supposedly Culbertson's locality. PU loc 1015F2a. W. J. Sinclair, August 14, 1920	VPPU.012592 *Subhyracodon* sp.
12593	*Mesohippus* sp.	Palate. West side of buttes south of 71 Table, supposedly Culbertson's locality. PU loc 1015F2a. H. R. Wanless, August 14, 1920	VPPU.012593 *Mesohippus bairdi*
12594	*Leptauchenia* sp.	Skull and jaws from the white ash zone of the *Leptauchenia* beds. Same horizon as specimen 12579. South side of Sheep Mountain, 1/8 mile south of Stony Pass between western pinnacles and main table. W. J. Sinclair, August 18, 1920	VPPU.012594 *Leptauchenia* sp.
12595	Coprolites	From upper part of *Titanotherium* beds, east branch of Indian Creek, adjoining Hart Table southwest of Taylor Ranch camp. PU loc 1015A1, upper part.	VPPU.012595 *Carnivore coprolites*
12596	Coprolites	From Spring Creek pocket, PU loc 1015B2a. W. J. Sinclair, August 11, 1920	No specimen in collection
12597	Coprolites & mouse-chewed bones	Nodular layer of Entelodon Peak area, PU loc 1015A2a. Indian Creek pocket. W. J. Sinclair	VPPU.012597 Carnivore coprolites, rodentchewed bones
12601	*Palaeolagus* sp.	From the red turtle-*Oreodon* layer, head of the east branch of Indian Creek. PU loc 1015A2a, lower *Oreodon* beds. H. R. Wanless, July 15 & 16, 1920. Found with *Leptomeryx* and *Palaeolagus* jaws, specimens 12527 and 12602.	VPPU.012601 *Palaeolagus haydeni*
12602	*Palaeolagus* sp.	From the red turtle-*Oreodon* layer, head of the east branch of Indian Creek. PU loc 1015A2a, lower *Oreodon* beds. H. R. Wanless, July 15 & 16, 1920. Found with *Leptomeryx* and *Palaeolagus* jaws, specimens 12527 and 12601.	VPPU.012602 *Palaeolagus haydeni*

Index

Printed in the USA
CPSIA information can be obtained
at www.ICGtesting.com
LVHW070409221123
764620LV00008B/543